EVERYDAY LIFESTYLES AND SUSTAINABILITY

The impact of humanity on the earth overshoots the earth's bio-capacity to supply humanity's needs, meaning that people are living off earth's capital rather than its income. However, not all countries are equal and this book explores why apparently similar patterns of daily living can lead to larger and smaller environmental impacts.

The contributors describe daily life in many different places in the world and then calculate the environmental impact of these ways of living from the perspective of ecological and carbon footprints. This leads to comparison and discussion of what living within the limits of the planet might mean. Current footprints for countries are derived from national statistics and these hide the variety of impacts made by individual people and the choices they make in their daily lives. This book takes a 'bottom-up' approach by calculating the footprints of daily living. The purpose is to show that small changes in behaviour now could avoid some very challenging problems in the future.

Offering a global perspective on the question of sustainable living, this book will be of great interest to anyone with a concern for the future, as well as students and researchers in environmental studies, human geography and development studies.

Fabricio Chicca is a senior lecturer in the School of Architecture, Victoria University of Wellington, New Zealand.

Brenda Vale is a professorial research fellow in the School of Architecture, Victoria University of Wellington, New Zealand.

Robert Vale is a professorial research fellow in the School of Architecture, Victoria University of Wellington, New Zealand.

EVERYDAY LIFESTYLES AND SUSTAINABILITY

The Environmental Impact of Doing the Same Things Differently

Edited by Fabricio Chicca, Brenda Vale and Robert Vale

First published 2018
by Routledge
2 Park Square, Milton Park, Abingdon, Oxon OX14 4RN

and by Routledge
711 Third Avenue, New York, NY 10017

Routledge is an imprint of the Taylor & Francis Group, an informa business

© 2018 selection and editorial matter, Fabricio Chicca, Brenda Vale and Robert Vale; individual chapters, the contributors

The right of Fabricio Chicca, Brenda Vale and Robert Vale to be identified as the authors of the editorial material, and of the authors for their individual chapters, has been asserted in accordance with sections 77 and 78 of the Copyright, Designs and Patents Act 1988.

All rights reserved. No part of this book may be reprinted or reproduced or utilised in any form or by any electronic, mechanical, or other means, now known or hereafter invented, including photocopying and recording, or in any information storage or retrieval system, without permission in writing from the publishers.

Trademark notice: Product or corporate names may be trademarks or registered trademarks, and are used only for identification and explanation without intent to infringe.

British Library Cataloguing-in-Publication Data
A catalogue record for this book is available from the British Library

Library of Congress Cataloging-in-Publication Data
A catalog record has been requested for this book

ISBN: 978-1-138-69387-6 (hbk)
ISBN: 978-1-138-69390-6 (pbk)
ISBN: 978-1-315-52913-4 (ebk)

Typeset in Bembo
by Deanta Global Publishing Services, Chennai, India

 Printed in the United Kingdom by Henry Ling Limited

CONTENTS

List of figures vii
List of tables viii
Acknowledgements x
List of contributors xi

1 Introduction 1
  Fabricio Chicca, Brenda Vale and Robert Vale

2 Africa 14
  Fabricio Chicca, Dashakti Reddy and Brenda Vale

3 Asia 35
  Fabricio Chicca, Jacqueline McIntosh, Yukiko Kuboshima, Jestin Nordin, Tri Harso Karyono and Brenda Vale

4 Europe 63
  Adele Leah, Hannu I. Heikkinen, Han Thuc Tran, Ludwig Geisenberger, Stefan Opfermann, Robert Vale and Brenda Vale

5 North America 76
  Fabricio Chicca, Silvio Marco Costantini, Abbie McKoy, Ana Paula Pagotto and Brenda Vale

6 Oceania 93
  Fabricio Chicca and Brenda Vale

7	South America *Emilio Garcia, Fabricio Chicca and Brenda Vale*	111
8	Calculating the ecological footprints of the stories *Fabricio Chicca, Sarah Nabyl-Calliou, Robert Vale and Brenda Vale*	126
9	Comparing the footprints *Fabricio Chicca, Robert Vale and Brenda Vale*	138
10	Food *Fabricio Chicca, Robert Vale and Brenda Vale*	152
11	Travel *Robert Vale, Fabricio Chicca, and Brenda Vale*	167
12	Energy *Brenda Vale, Robert Vale and Fabricio Chicca*	188
13	Dwelling *Brenda Vale, Robert Vale and Fabricio Chicca*	208
14	Consumer goods *Brenda Vale, Robert Vale and Fabricio Chicca*	218
15	Conclusions *Brenda Vale, Robert Vale and Fabricio Chicca*	227
Index		*233*

FIGURES

1.1	Break down of the ecological footprint of a citizen of Cardiff, UK	8
2.1	Typical shelters, PoC3, UNMISS, Author, June 2016	30
2.2	Typical cooking facilities, PoC3 UNMISS, Author, June 2016	31
3.1	Typical dinner (left) and lunch box (right)	48
9.1	Location of the countries which are in the Top 20 ranking of the Index	139
9.2	HDI, average income, density (p/ha) and index for batches of 30 countries, grouped by index	141
10.1	A diagram depicting how food miles may be caused	153
10.2	Food transport emissions by vehicles in kg per tonne-km	156
10.3	The environmental impact for food in gha per person for a year	161
10.4	The environmental impact for food in CO_2 and ecological footprint	162
10.5	The environmental impact for food in gha per person and the annual consumption of animal products including all kinds of meats, dairy products and eggs [expressed in kg per person]	163

TABLES

1.1	Locations of the stories	2
2.1	The family's meals for a week in April 2016	25
2.2	Family meals for a week in April 2016	28
2.3	The family's meals for a week in April 2016	32
3.1	Asian sub-regions, population, and story locations	36
3.2	Summary of EFs and urbanisation of story countries	36
4.1	Typical weekly menu	73
4.2	Typical monthly car usage and CO_2 emissions	73
6.1	Food	98
7.1	EFs of South American nations	112
8.1	Carbon dioxide sequestration potential in three climates from IPCC (2007)	130
8.2	Ecological footprints for a range of electricity generation options	131
8.3	Coal equivalents of fuels used directly	132
8.4	Energy mix for generation of electricity in Finland	133
8.5	Average CO_2 emissions for electricity generation sources	133
8.6	Embodied energy of houses	134
8.7	Carbon dioxide sequestration rates for different climates	135
8.8	CO_2 emissions calculated for each family's appliances	135
9.1	Conversion factors/indices and densities of countries with the Top 10 indices	140
9.2	Index for each country in the stories	142
9.3	EF of each family as calculated here in gha per person	143
9.4	Comparison of EF as calculated here and average EF in gha from GFN data	144
9.5	CO_2 per capita emissions for each family compared with EFs based on national indices	148

List of tables ix

9.6	Breakdown of EFs in stories using national indices gha/person	149
9.7	Breakdown of CO_2 emissions per capita	150
10.1	The amount of energy that could be produced from different food types compared with one unit of energy from meat	157
10.2	Milk compared with various plant crops	158
10.3	The energy efficiency (energy in/energy out) of crops, adapted from Pimentel (2009)	159
11.1	Emissions per passenger-kilometre for bicycles, car and bus	169
11.2	CO_2 emissions in grams per passenger-kilometre for a range of urban and long-distance transport modes in the United States	170
11.3	CO_2 equivalent figures per passenger-kilometre	171
11.4	CO_2 emissions for freight in grams per tonne-kilometre	173
11.5	CO_2 emissions related to transport for each of the stories	180
11.6	Carbon dioxide emissions for transport kg per person, annual distance travelled per person per year and emissions in grams per person-km per year	183
12.1	What you can do with the energy in a barrel of oil	190
12.2	Emissions from direct burning of selected fossil fuels	193
12.3	Comparison of emissions from generation of electricity	197
12.4	Sources of energy in the stories and energy EF in gha	198
12.5	Energy that can be generated from a hectare in southern Germany	200
12.6	Comparison of carbon and ecological footprints per person	201
13.1	Dwellings from the stories ordered by size	210
13.2	Houses ranked by occupancy, location, and whether they are detached or conjoined in some way	213
13.3	Ranking of houses by EF, overall area and area/person	216
14.1	Access to a working computer at home	222
14.2	Appliance ownership from the stories	223
14.3	Comparison of CO_2 emissions and EFs for consumer goods	224
15.1	EFs and well-being in Wellington, New Zealand	229
15.2	Comparison of EF, Happiness Index and GDP	230

ACKNOWLEDGEMENTS

The authors would first like to extend very grateful appreciation to all the families who offered to take part in this book by describing their lives and to the people who researched and wrote some of these. Special thanks are given to Sarah Nabyl-Caillou for all her work on the ecological footprint calculations, and to Nilesh Bakshi for his help in dealing with domestic appliances.

CONTRIBUTORS

Fabricio Chicca is a senior lecturer in the School of Architecture, Victoria University of Wellington, New Zealand.

Silvio Marco Costantini is a software engineer at NP6 in Bordeaux, France.

Emilio Garcia is an architect and urban designer. Since 2013 he has been working as a lecturer in sustainability at the School of Architecture and Planning in the University of Auckland, New Zealand.

Ludwig Geisenberger is retired from teaching physics and chemistry in Germany.

Hannu I. Heikkinen is professor of cultural anthropology at the University of Oulu in Finland.

Tri Harso Karyono is a professor of architecture, specialising in sustainability, at Tanri Abeng University, Jakarta, Indonesia.

Yukiko Kuboshima is a PhD candidate in the School of Architecture, Victoria University of Wellington, New Zealand.

Adele Leah is a British qualified architect and a senior lecturer in the School of Architecture at Victoria University of Wellington, New Zealand.

Jacqueline McIntosh is an active researcher in sustainable housing and a senior lecturer in the School of Architecture, Victoria University of Wellington, New Zealand.

Abbie McKoy is a placemaking and community development professional, currently working on the city centre revitalisation project at Porirua City Council.

Sarah Nabyl-Calliou is a graduate student in engineering and architecture at ENPTE (Ecole Nationale des Travaux Publics de l'Etat) and ENSAL (Ecole Nationale Supérieure d'Architecture de Lyon), Lyon, France.

Jestin Nordin teaches architecture subjects and studios at the School of Housing, Building, and Planning, University Sains Malaysia, Penang, Malaysia.

Stefan Opfermann is an inspector for an ecological certification organisation in Germany.

Ana Paula Pagotto is a speech therapist graduated by the Escola Paulista de Medicina and postgraduated in Communication and Marketing by the IBMEC Brazil.

Dashakti Reddy is a humanitarian aid worker, focusing on women's protection and empowerment and has worked in Liberia, South Sudan and Iraq.

Han Thuc Tran is a postdoctoral researcher at the University of Oulu, Finland.

Brenda and Robert Vale are professorial research fellows in the School of Architecture, Victoria University of Wellington, New Zealand. List of IllustrationsTables

1
INTRODUCTION

Fabricio Chicca, Brenda Vale and Robert Vale

What this book is about

For the first time in human history, the majority of the human population lives in urban rather than rural areas (United Nations, 2015). This book asks whether or not we should be worried about this.

To investigate our impact on the environment, we present stories of how people live their daily lives in 23 different places from the 6 inhabited continents. The individuals in these stories have been given fictitious names to preserve their anonymity. Only those who have written up the stories have been named. Table 1.1 shows the locations of the stories and whether they are broadly rural or urban.

Overall, 15 of the locations could be described as urban in character and the remaining 8 as rural.

Definition of urban areas

There is a big difference between living in a very large city and in a small town, although both are urban in character. Not every country defines an urban area in the same way. In the US census, an urban area has in excess of 50,000 people and an urban cluster has from 2,500 to 50,000 people, with everything else being considered rural (US Census Bureau, 2016). In contrast, the 2011 census in India had four urban categories (statutory towns, census towns, urban agglomerations and urban outgrowths). A town is categorised as a settlement with a minimum population of 5,000 people, a density of not less than 400 people per square kilometre and at least 75 per cent of the male population of working age not engaged in agriculture (Census India, 2011). This can be compared with the 6,400 people per square kilometre of São Paulo (Cox, 2012), showing how varied a definition of "urban" can be. When it comes to whether a place is urban or rural, it is more a matter of

2 Fabricio Chicca, Brenda Vale and Robert Vale

TABLE 1.1 Locations of the stories

Continent	Country	Place	City	Town	Rural/village
Africa	Morocco	Marrakesh	√		
	Mozambique	Inhassoro			√
	South Africa	Johannesburg	√		
	South Sudan	Juba*	√		
Asia	India	Khajuraho			√
	Indonesia	Kendeng Mountains			√
	Japan	Nagoya	√		
	Malaysia	Penang	√		
	Mongolia	Ulaanbaatur	√		
	Myanmar	Lake Inle			√
Europe	Finland	Oulu	√		
	Germany	Eschenlohe			√
	United Kingdom	Newton-le-Willows		√	
North America	Canada	Toronto	√		
	Cuba	Havana	√		
	United States	St. Tammany Parish			√
	United States	Celebration		√	
Oceania	Australia	Sydney	√		
	New Zealand	Near Stratford			√
	Tonga	Nuku'alofa	√		
South America	Argentina	Tucumán	√		
	Brazil	Dourados			√
	Brazil	São Paulo	√		

*There are two stories from Juba

common sense than an agreed definition: if it is dominated by buildings it is urban, whereas if fields and trees dominate it is rural. This is the basis of the definitions in Table 1.1 for the stories in this book.

The Indian definition implies that you are urbanised if you are no longer engaged in agricultural activities. This change from rural to urban employment is happening in the name of urbanisation, but should we be worried about this natural outcome of the industrialisation of agriculture? Pelletier et al., (2011) show that modern systems for providing food rely largely on inputs of non-renewable energy, from the farm to retailing. As people move to the city, fossil fuel energy has replaced the human and animal energy formerly used. Antonini and Argilès-Bosch (2017) suggest that the energy consumption of farming will rise even further with increasing mechanisation. Given climate change, putting more fossil fuel energy into farming is untenable. It is also untenable because oil is a finite resource. Whatever economists might argue, we cannot continue to extract oil from the earth forever.

As we shall see later in this book, this continuing and increasing use of oil to feed us should be a cause for concern, given that food is the major component of all personal ecological footprints. Ecological footprints, or EFs, are a way of measuring the environmental impact of a person, an activity, a product or a country based on the area of land that would be needed to meet the demand in a sustainable manner (Wackernagel and Rees, 1996). The EF, which is measured in "global hectares" (gha), is defined in greater detail later in this chapter.

Urban living

Living in cities appears to consume more resources overall than living in rural areas, even in the developed world. In 2001, the EF of a citizen of the city of Cardiff was 5.6 gha/person while that of someone living in rural Gwynedd was 5.25 gha/person (Garcia and Vale, 2017: table 8.3). Hubacek et al., (2009) compared the 2001 average EF for China of 1.8 gha/person with that of Beijing, which was 4.99 gha/person.

If we live in cities, especially at high densities, we cover less land with buildings and roads, leaving more land for growing crops. However, these crops have to be processed into a form in which they can be brought into the cities to be sold and used, so energy is put into transforming, packaging and transporting them. All the wastes generated in the city have to be shipped out and disposed of somewhere. In the past, systems were worked out to deal with these problems. In ancient Athens, sewage and storm water were collected in a basin outside the city and directed in brick-lined conduits to fields and orchards as fertiliser to produce the food that the people of Athens could eat to produce the sewage, and so on *ad infinitum* (Antoniou et al., 2014: 100). Such systems work where populations are reasonably small. In the fifth century BCE, Athens was certainly not a village; it had a maximum population of 140,000, with around 40,000 male citizens and 40,000 slaves (Educational Resources, n.d.). In eighteenth-century London, the city was still small enough that human sewage could be collected and stored at Dung Wharf for shipping downriver to market gardens along the River Thames, the barges returning with food (Parks and Gardens UK, n.d.). By 1760, the population of London had risen to 760,000 (Old Bailey Proceedings, 2015). Like ancient Athens, this is a small city to modern eyes. This explains why in modern agriculture natural fertilisers have been replaced with chemical fertilisers, which take energy to manufacture and which use finite resources rather than recycled wastes. If (or is it when?) these finite resources run out, the food will run out too.

By 2016, there were 512 cities across the world with a million or more inhabitants (UN ECOSOQ, 2017: 9). Many of the stories in this book are of people who live in cities. One purpose of looking at all these stories is to understand whether it is possible to live in a city and still have a low environmental impact.

Why we live in cities

Why do we live in cities, and why is living in cities on the increase? Essentially, cities are about trade (Mortazavi, 2010: 126), emerging as marketplaces for the products of the surrounding agricultural areas (Jacobs, 1969: 38). For the majority of history, people walked from the surrounding countryside to the nearest settlement with a market where they could trade their produce. Trade was originally to do with swapping what you could grow for what you could not, leading to wealth. It is this lure of an easy path to riches that makes living in cities, where, according to the traditional story of Dick Whittington, the streets are paved with gold, so attractive.

Even before nineteenth-century industrialisation, the lives of those in urban areas were very different from those of their rural contemporaries. People in cities went to work to earn money to buy the food grown by the rural population. This distance between doing things for yourself and earning the money to buy the same things leads to a disassociation. If you grow something, you know the time and effort required to produce it. If you grow a potato, you have dug and planted and hoed and harvested it yourself. If you buy a potato, all that effort is hidden and you only value the potato in terms of what it costs you. This could lead to bought goods being less appreciated, producing what has been called the "throwaway consumer society" (Vince, 2012). If you have spent several weeks of spare time knitting a sweater, you are unlikely to throw it out until it becomes really unwearable, because you know what went into making it.

As Table 1.1 shows, the majority of the stories in this book are set in urban rather than rural locations. However, whether urban or rural, for most people the daily pattern of living is very similar. You get up, go to work and come home at the end of the day. The work day for children is spent in school. If you are poor, you may not be able to afford to send your children to school, or you may need them to work to help raise the family income. This book, through comparing daily life for 24 families with different levels of wealth and expectations, tries to work out the environmental impact of these typical days to see what the big environmental impacts in these daily patterns are, and how these impacts might be able to be reduced.

Urbanisation and food

We all need food. In Ancient Greece, as urbanisation spread, food supply became a big social concern. In c600 BCE, the city of Corinth forbade its citizens to leave their land because this would have put food production and other resources—leather and wool, for example—at risk (Gutkind, 1969: 472). Cities allowed a certain division of labour, so those not engaged in agriculture could become traders and producers of other goods (Gutkind, 1969: 467). This division of labour increased during the time of the Roman Empire, probably the first civilisation with a strong concept of urbanisation, using the booty from its conquests to create an extensive built environment, including a road network, which further advanced trade and other economic activities (Kay, 2014: chapter 9). Before Rome became a populous city, it was relatively hard to find a citizen who did not have an underlying relationship with the countryside. Many people used to work in the countryside for part of the year, or even daily, and this provided pressure to constrain the city and keep it moderately small due to the need to be close to its food supply (Montillo, 1956: 25). People could walk out to the fields to help with food production at critical times, like the harvest. Once urbanisation was spread throughout the Empire and the cities grew in size, trade became essential to maintain the increasingly indirect relationship between consumption and production. The urban citizen became a consumer instead of an agricultural producer. Because the people in Rome lost the sense of the effort involved in the production of foodstuffs, they could easily consume more food than necessary, providing they could earn the money to pay for it. Wealthy urban Romans enjoyed the tongues of flamingos and honey-roasted dormice rolled in poppy seeds (Allan, 2005: 70). Even "soldiers serving on Hadrian's Wall ... list their foods as ... spice, goat's milk, salt, young pig, ham, corn, venison and flour ... vintage wine, Celtic beer, ordinary wine, fish sauce and pork fat ..." (Renfrew, 2004: 7).

With urbanisation, the connection between people and the production of food has been lost. Rees and Moore (2013: 11) suggest that cities are now human feedlots. A feedlot is where livestock are brought to be fattened—it contains one species in a small space, the food all has to be brought in and all the wastes have to be disposed of, just like a modern city.

Why worry about living in cities?

Although people living in cities might forget how dependent they are on land to grow their food and timber, wool and cotton, this does not yet explain why we should be worried about living in cities, since the earth has plenty of land. If the entire world population stood side by side, each person occupying 1 square metre of land, it would cover an area slightly smaller than the small state of Delaware in the United States (Snapzu, n.d.). The problem is that we all consume much more land than we would if we were just standing on our own square metre. This is where the EF comes in, as it measures human impact in terms of the amount of land needed to supply each person with all the resources they consume in their daily living.

The concept of the EF

In their 1996 book *Our Ecological Footprint*, Mathis Wackernagel and William Rees proposed a new interpretation of the old idea that land was the key factor in the sustained development of a society. They proposed that human impact could be visualised as the impression of a person's footprint on the eco-systems that support us all. In the case of the EF, that impression represents the resources removed from the eco-system to supply what each person takes to live their life. What the EF aims to capture is the impact of all the processes that go to make up a product or service that each person consumes. The calculation assumes that the land has to provide everything, on the basis that in a world where all the finite resources have been consumed, this will be the reality. In the case of copper, for example, the calculation would include all the energy that goes into extracting and making it, and whether the energy is renewable (in which case, determining how much energy can be produced from a hectare of land) or fossil fuel-based (in which case, determining how much land is needed to absorb the carbon dioxide emitted from burning the fuel). To this is added the land degraded by the mining process, as well as the land required for the energy that goes into transporting the product to where it is used, and also into retailing it; the products that go to make up the chemicals used to extract the copper; and an allowance for the energy needed to recycle the copper when the product comes to the end of its useful life. Thus a simple visualisation of human impact—the EF—becomes highly complex when it comes to calculating the impact of materials that we expect to use.

The EF is easier to visualise when it comes to food. If it takes one potato plant in a mulched raised bed to produce one kilogram of potatoes per year (Pleasant, 2015) and the recommended planting spacings for potato plants are 300 millimetres apart in rows of 750 millimetres apart (Hall, 2016), then in a plot 10 metres × 10 metres, there will be 13 rows with 33 plants in each, so I might expect 429 kilograms of potatoes from that 100 square metres of land. The average person in the United Kingdom eats 429 grams of potatoes each week (Statista, 2017), so our plot of potatoes will feed 19 people in the United Kingdom each year and the footprint of a UK person's annual potato consumption is a piece of land roughly 5 square metres in area. This takes no account of weather, pestilence or other problems that might affect the yield. EFs deal in averages gleaned from existing data, so in a sense, they are always out of date. Potato consumption is falling in the United Kingdom (Statista, 2017), so arguably our plot could feed more people, but then people will be substituting other food for the potatoes they previously ate, and these other foods might need more land than potatoes do. Another problem is that the productivity of the soils will be different within each country. This means that if the footprints of food and other products are to be comparable between countries, we have to take land productivity into account. This is something we shall return to in defining the global hectare, which is the unit used in EFs.

Definition of the EF

The EF is the area of biologically productive land and water an individual, population, or activity requires to produce all the resources it consumes and to absorb the wastes it generates in a sustainable way (Global Footprint Network, 2012). You can measure the EF of a person, a product, a city or a whole country. Wackernagel and Rees (1996: 94–95) give the example of the Netherlands, with an EF of nearly the same area as the whole of France, showing that the Dutch use of resources outstrips the land available within their borders, even given their famed ability to reclaim land from the sea. The French, not unnaturally, want to keep their land for themselves, so where can the Dutch turn to obtain what they want? What the EF does is to reveal which countries are dependent on land beyond their own borders for the resources they consume, often an area of land far larger than the land they occupy.

The problem with this definition is that not all resources can be secured on a sustainable basis. Obviously, you can grow crops, but you cannot grow light bulbs, so the EF is not truly measuring whether a particular lifestyle is sustainable. To overcome this problem, the EF looks at the land needed to produce and recycle the resource as well as the land needed to grow the energy to do this.

The ultimate aim of the EF is to see whether it would be possible to extract all the resources used by people on earth from the land (and water) on a sustainable basis. As the population rises and the area of biologically productive land remains the same, each person's share of all the available land from which they would need to supply all their needs—the "fair earth share"—becomes smaller. The estimate in 2013, with 12 billion hectares of productive land and 7 billion people, was that the fair share was 1.7 hectares per person (Rees and Moore, 2013: 15).

Wackernagel and Rees (1996: 90) state that EF can be used as a way of comparing the impact on the environment of different people and societies. This is how the EF is used in this book. Our stories are not sufficiently detailed to give the complete EF of each person but they can be used to calculate and compare important aspects that each person, in a sense, has under their control, things over which we have a degree of choice. Figure 1.1 sets out how the EF of a citizen of Cardiff in Wales (Collins et al., 2015) breaks down into different categories and shows which ones an individual has some control over (Individual) and those they cannot control (Collective). The part labelled "Grey" is what would be required to have a footprint reasonably close to the fair share, showing how the total (Grey plus White) would have to be reduced if each part were reduced by the same percentage.

The most surprising aspect of Figure 1.1 is that the largest part of the EF is food. As we shall see from the stories, if you are very poor, almost the whole of your EF is the impact of the food you eat, and food also powers your travel. This book only deals with the aspects of the EF that are under the control of the individual. Each story will be analysed to see the impact of food, domestic energy use, household travel, consumer goods and the house, and the results compared.

Bio-capacity

The concept of EF can also be used to determine how many people the earth can support for a given style of life. A higher material quality of life means a greater environmental impact because having more material goods means using more resources to make them. Cities in the past were small because there was a link between the availability of resources, particularly food, and the size of the city. In Ancient Greece, when the population increased, the city had to organise itself to send some of its citizens off to found a new colony. There is no longer room for new colonies. What the EF demonstrates starkly is that the more the human population grows, the fewer are the resources available to be shared between each person.

Human beings are a very successful species, at least in terms of numbers. As of August 2017, the human population was over 7.5 billion (Worldometers, 2017). In the 1950s, the world population was below 3 billion and we could all have lived the lifestyle of 1950s Europe, while in 2012, we could all live like people in Cuba (Vale and Vale, 2013: 320). As the population rises in a finite world, there are fewer

8 Fabricio Chicca, Brenda Vale and Robert Vale

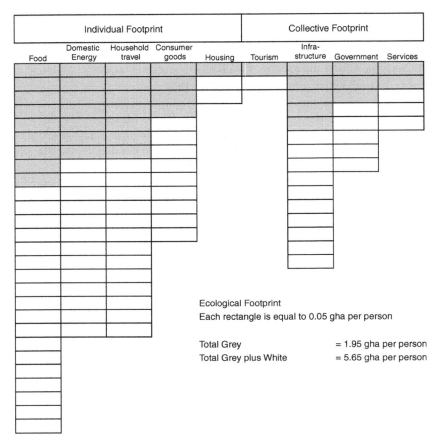

FIGURE 1.1 Breakdown of the ecological footprint of a citizen of Cardiff, United Kingdom.

resources to go around. The United Nations is predicting that in 2050 the global population will be between 9.4 and 10.2 billion, and that by 2100 it will have risen to between 9.6 and 13.2 billion (UN DESA, 2017: 12). What the EF tells us is that this is going to be a big problem.

The reason we can live much more resource-intensive lifestyles than those that would result from living sustainably off the land, and relying only on solar energy, is that most people make use of the stored resource of fossil fuels laid down during the age of the dinosaurs. This has led to problems in terms of climate change and the need to curb carbon emissions. The use of this "buried treasure" of oil, gas and coal means that there is no need to make use of renewable resources, and if we did have to use them, they might not be sufficient to support our current wants. The date when humanity's demand in a given year for ecological resources and services exceeds what the Earth can regenerate in that year is known as Earth Overshoot Day (Earth Overshoot Day, 2017). In 2015, demand outstripped supply in mid-August

(WWF, 2016). In 2017, it happened on 2 August (Frangoul, 2017). These ecological resources and services are known as bio-capacity. In 2013, only 55 countries lived within their bio-capacity, while 136 did not (GFN, 2017). Countries that exceed their bio-capacity have to trade for the necessary ecological resources and services.

In the past, local cultures and traditions arose because the ecological resources and services available locally were different. Modern economic globalisation has arisen from the fact that the ideologies of the free market and trade are now commonplace throughout the world (Anon, 2002). People with different backgrounds and cultures in different parts of the world can now compare their apparent living standards with others. Unsurprisingly, as people start to do this, some standards are perceived as better. The assumption is that the best way to achieve these standards is to adopt the work and consumption patterns found in the wealthier regions, which helps the wealthier regions become even wealthier through expanding the markets for the products they make. This conceptual structure, which surreptitiously invites people to join this modern way of living while repudiating their old traditions, may have the same importance that Adam Smith had for the industrial revolution and the rise of capitalism. What the EF reveals is that this may not be possible if the world is to survive into the long-term future. The future requires a reduction in resource-intensive ways of living, not their spread.

How is the EF measured?

The unit of measurement for the EF is the global hectare (gha), which represents an averagely productive hectare of the biologically productive land and water a population needs to produce the renewable resources it uses and to absorb the wastes it creates. The renewable resources are crops, animal products, timber and fish. Also taken into account is the land related to the consumption of energy and the loss of land to built-up areas (usually called degraded land). These are all calculated on an annual basis. Obviously, some land is going to be more productive, either because of the climate or the type of soil, or both. The global hectare accounts for these differences by using world average productivity for each type of productive land. An equivalence factor (gha/ha) is used to translate local hectares into global hectares (Schaefer et al., 2006). These equivalence factors apply to all countries but vary from year to year, not least because of changing weather patterns. We shall return to this in Chapter 8 in the discussion of how the EFs in this book were calculated.

Wildlife

People are not the only inhabitants of planet earth. Other creatures also need the resources and services of eco-systems for their own survival. How much land should be left aside for biodiversity, representing the 30 million species other than human beings that live on earth (Wackernagel et al., 1999)? In 1987, almost 4 per cent of the world's land area was protected or managed to conserve species (WCED, 1987: 146). The World Commission on Environment and Development recommended that this protected area needed to be tripled (WCED, 1987: 166). This has led to those engaged in EF accounting setting aside 12 per cent of all productive land and water for other species in their calculations (Wackernagel et al., 1999).

Carbon footprint

Energy is a component of the EF, but with climate change and the need to reduce greenhouse gas (GHG) emissions, carbon footprint (CF) is now often discussed separately from EF. What is crucial is that EF and CF are not measuring the same thing and should never be confused. The CF is a measure of the GHG emissions resulting from using something like a domestic appliance or that are emitted during a service, such as taking a flight. The CF is measured in kg or tonnes of carbon dioxide (CO_2) or CO_2 equivalent. This can then be converted into the gha needed to absorb or sequester this quantity of CO_2 to prevent it going into the atmosphere and adding to global warming. CF renders something hard to visualise—a tonne of invisible gas—into something much easier—an area of land. A problem with this approach is that if we sequester the GHG emissions in plants or trees and if these rot or are burnt, the GHG will return to the atmosphere, so to absorb GHG emissions more land needs to be set aside for the cyclical process of growth and decay.

Top-down and bottom-up

Before we leave this general discussion of EF, it is important to note that it is generally calculated in one of two ways—top-down and bottom-up. National EFs can be calculated from national statistics and allow for imports and exports—how much food a country grows and consumes, how much energy it generates and consumes, and how much energy and resources go into the goods it produces and trades (Wackernagel et al., 1999). This can then be compared with the area of biologically productive land it has available, accounting for the built-up land occupied by human settlement.

The other way to measure EF is a bottom-up process, and this forms the basis of personal EF online calculators (GFN, n.d.; WWF, n.d.). The aim is first to find out what you do, how far you travel, how much energy you use, how much stuff you have and what you eat, and then to calculate the impact of this based on national data. This is the approach taken in this book by asking people how they live and what they do on a daily basis. It is much harder to capture all information using this approach, but the aim with the stories in this book is to calculate all EF impacts on a similar basis so that they can be compared (see Chapter 9) and the impact of how we live can be revealed.

What does the EF teach us?

Given all the problems of collecting the information and calculating an EF, is it really useful? EFs are always out of date as they are based on past data. The real advantage of knowing the EF of a person, or even a product, is that similar things can be compared and lessons learnt about what we might have to do if we are truly serious about not living beyond our means by constantly overshooting the earth's sustainably produced resources.

Three centuries ago, people around the world shared a roughly similar quality of life and had roughly similar environmental impacts. The industrial revolution not only increased the disparity between people in the same area because of increased differences in earnings but also created a major difference between the environmental impacts of people in different parts of the world. Today, a simple way of living, without engaging in consumerism, is generally perceived as backward and old-fashioned. The perception that engaging in the consumerist way of life is the natural progression from the simplicity of the past has had dramatic consequences for the environment. Environmental consciousness has been increasing (Kachur et al., 2010), which has led to attempts to make the modern consumerist lifestyle more sustainable through the environmental labelling of consumer products. These labels are another facet of consumerism, and it is sometimes now considered fashionable to wear, live in or drive something "environmentally friendly". However, labelling a product is not the same as labelling the impact of a whole lifestyle. The EF reveals whether labelling is really sufficient to reduce the impact or whether, to make a difference, we have to face up to the idea of not having the consumer products.

This book sets out to promote discussion about the impact of a whole lifestyle through comparing and measuring the same basic activities in societies with very different environmental impacts, using the EF. People may live in different climates and have different governments but they still tend to live in similar ways. They engage in economic activities to earn money, they raise children, they celebrate, they worship, they even play similar sports, but despite these very similar habits, their environmental impacts are remarkably different.

Through the stories in this book of similar behaviours in different societies, the objective is to explain why some cultures have lower environmental impacts than others. Scenarios will emerge of ways of living with a lower environmental impact. The aim is to make these scenarios more than just numbers, through describing what a low-impact world might look like based on real examples. The stories have deliberately been drawn from low- and high-EF societies and from all urbanised continents.

The book starts with the stories, arranged alphabetically by continent and country. Following the stories from six continents, Chapter 8 deals with how the EFs were calculated. Chapter 9 looks at the results. Chapters 10–14 discuss the aspects of the EF that relate to daily life—food, energy, travel, the dwelling and consumer goods. Chapter 15 ends by discussing what the stories and the EFs tell us about living within the resources and eco-systems services of our only planet, the home of humanity.

References

Allan T (2005) *Life, Myth, and Art in Ancient Rome*, Los Angeles: Getty Publications.
Anon (2002) Time to rethink everything, *New Scientist*, 174(2340), pp. 29–52.

Antonini C and Argilès-Bosch J M (2017) Productivity and environmental costs from intensification of farming. A panel data analysis across EU regions, *Journal of Cleaner Production*, 140(2), pp. 796–803.

Antoniou G P, Lyberatos G, Kantetaki E I, Kaiafa A, Voudouris K and Angelakis A N (2014) History of urban wastewater and stormwater sanitation technologies in Hellas, in Angelakis A N and Rose J B (eds) *Evolution of Sanitation and Wastewater Technologies Through the Centuries*, London: IWA Publishing, pp. 99–146.

Census India (2011) Provisional population totals: urban agglomerations and cities, available at http://censusindia.gov.in/2011-prov-results/paper2/data_files/India2/1.%20Data%20Highlight.pdf, accessed 29 July 2017.

Collins A, Flynn A and Netherwood A (2005) *Reducing Cardiff's Ecological Footprint*, Cardiff: WWF Cymru, Sustainable Development Unit Cardiff Council and The Centre for Business, Relationships, Accountability, Sustainability and Society.

Cox W (2012) Evolving urban form: São Paulo, available at http://www.newgeography.com/content/003054-evolving-urban-form-s%C3%A3o-paulo, accessed 29 July 2017.

Earth Overshoot Day (2017) About Earth Overshoot Day, available at http://www.overshootday.org/about-earth-overshoot-day/, accessed 4 August 2017.

Educational Resources (n.d.) *Two faces of Greece: Athens and Sparta*, available at https://www.pbs.org/empires/thegreeks/educational/lesson1.html, accessed 7 August 2017.

Frangoul A (2017) *It's Earth Overshoot Day: We've used more resources than nature can regenerate in 2017*, available at https://www.cnbc.com/2017/08/02/its-earth-overshoot-day-weve-used-more-resources-than-nature-can-regenerate-in-2017.html, accessed 4 August 2017.

Garcia E J and Vale B (2017) *Unravelling Sustainability and Resilience in the Built Environment*, London: Routledge.

Global Footprint Network (GFN) (n.d.) What is your ecological footprint, available at http://www.footprintcalculator.org/#!/signup, accessed 5 August 2017.

Global Footprint Network (GFN) (2012) *Trends*, available at http://www.footprintnetwork.org/en/index.php/GFN/page/trends, accessed 31 July 2012.

Global Footprint Network (GFN) (2017) Ecological wealth of nations, available at http://www.footprintnetwork.org/content/documents/ecological_footprint_nations/, accessed 4 August 2017.

Gutkind E (1969) *Urban Development in Southern Europe: Italy and Greece*, Toronto: Collier Macmillian.

Hall D (2016) *7 ways to grow potatoes*, available at https://www.rodalesorganiclife.com/garden/7-ways-grow-potatoes, accessed 3 August 2017.

Hubacek K, Guan D, Barrett J and Wiedmann T (2009) Environmental implications of urbanization and lifestyle change in China: Ecological and Water Footprints, *Journal of Cleaner Production* 17(14), pp.1241–1248.

Jacobs J (1969) *The Economy of Cities*, New York: Random House.

Kachur M, Bruck D and Hesedahl J (2010) Acoustical materials for a green world: The sustainable design transformation of the architectural acoustics industry, *The Journal of the Acoustical Society of America*, 127(3), p. 1723.

Kay P (2014) The creation of 'material complexity', in Kay P (ed.) *Rome's Economic Revolution*, Oxford Scholarship Online, DOI:10.1093/acprof:oso/9780199681549.001.0001.

Reisemann S (ed.) (2007) *Sustainable Metals Management: Securing Our Future – Steps Towards a Closed Loop Economy*, Dordrecht: Springer e-book, pp. 97–113.

Montillo P (1956) *Historia Antigua*, Buenos Aires: Buenos Aires Libre.

Mortazavi M (2010) From ancient to modern urbanization: Intermediary function of an urban society, *Journal of Historical Archaeology*, 15(1), 126–137.

Old Bailey Proceedings (2015) *A population history of London*, available at https://www.oldbaileyonline.org/static/Population-history-of-london.jsp#a1715-1760, accessed 27 July 2015.

Parks and Gardens UK (n.d.) *Night soil and other euphemisms…*, available at https://parksandgardensuk.wordpress.com/2015/07/11/night-soil-and-other-euphemisms/, accessed 27 July 2017.

Pelletier N, Audsley E, Brodt S. Garnett T, Henriksson P, Kendall A, Kramer K J, Murphy D, Nemecek T and Troell M (2011) Energy intensity of agriculture and food systems, *Annual Review of Environment and Resources* 36, pp. 223–246.

Pleasant B (2015) A simple way to get high yields of potatoes, available at https://www.growveg.com.au/guides/a-simple-way-to-get-high-yields-of-potatoes/, accessed 3 August 2017.

Rees W E and Moore J (2013) Ecological footprints and urbanization, in Vale R and Vale B (eds) *Living within a Fair Share Ecological Footprint*, London: Routledge, pp. 3–32.

Renfrew J (2004) *Roman Cookery: Recipes and History*, London: English Heritage.

Schaefer F, Luksch U, Steibach N, Cabeça and Hanauer J (2006) *Ecological Footprint and Biocapacity*, Luxembourg: European Communities.

Snapzu (n.d.) *How much room the entire world population would take up if standing side by side*, available at https://snapzu.com/geoleo/how-much-room-the-entire-world-population-would-take-up-if-standing-side-by-side, accessed 3 August 2017.

Statista (2017) *Average purchase per person per week of fresh potatoes in the United Kingdom (UK) from 2006 to 2015 (in grams)*, available at https://www.statista.com/statistics/284391/weekly-household-consumption-of-fresh-potatoes-in-the-united-kingdom-uk/, accessed 3 August 2017.

United Nations (2015) *Urban and Rural Population by Age and Sex, 1980-2015*, New York: United Nations.

United Nations Department of Economic and Social Affairs (UN DESA) (2017) *World Population Prospects: The 2017 Revision*, New York: United Nations.

United Nations Economic and Social Council (2017) Commission on Population and Development, available at https://digitallibrary.un.org/record/858729/files/E_CN.9_2017_5-EN.pdf, accessed 13 December 2017.

United States Census Bureau (2016) Urban and Rural, available at https://www.census.gov/geo/reference/urban-rural.html, accessed 29 July 2017.

Vale B and Vale R (2013) 'I wouldn't start from here…', in Vale R and Vale B (eds) *Living within a Fair Share Ecological Footprint*, London: Routledge, pp. 319–321.

Vince G (2012) *The high cost of our throwaway culture*, available at http://www.bbc.com/future/story/20121129-the-cost-of-our-throwaway-culture, accessed 28 July 2017.

Wackernagel M and Rees W E (1996) *Our Ecological Footprint*, Gabriola Island: New Society Publishing.

Wackernagel M, Onisto L, Bello P, Linares A, Falfan I, Garcia J, Guerrero A and Guerrero M (1999) National natural capital accounting with the ecological footprint concept, *Ecological Economics* 29, pp. 375–390.

World Wildlife Fund (WWF) (n.d.) *How big is your environmental footprint?*, available at http://footprint.wwf.org.uk/, accessed 5 August 2017.

World Wildlife Fund (WWF) (2016) Overshoot Day, available at http://www.worldwildlife.org/pages/overshoot-day, accessed 8 June 2016.

World Commission of Environment and Development (WCED) (1987) *Our Common Future*, Oxford: Oxford University Press.

Worldometers (2017) World population, available at http://www.worldometers.info/, accessed 4 August 2017.

2
AFRICA

Fabricio Chicca, Dashakti Reddy and Brenda Vale

Background

Brenda Vale

Stanley (1878/1890), in his writings about Africa, may not have invented the idea of the "dark continent" (Brantlinger, 1985), but he did popularise it. Since then, the idea of "emerging from the dark to the light" in the modern development of Africa and her peoples has become a useful metaphor when it comes to writing up academic research, from conserving biodiversity (Pimm, 2007) to examining changing attitudes to African American culture (Jones 2010). The fact that Africa is not yet as developed as the Western world means, from the viewpoint of ecological footprint (EF), the west owes Africa a big debt, but this is changing. In 1979, the average EF for Africa was only 0.7 gha/person, but by 2008 it had climbed to 1.4 gha/person (AfDB and WWF, 2012: 12). Although still under the fair share, this was almost equal to the bio-capacity of the continent of 1.5 gha/person (AfDB and WWF, 2012: 6). Africa is the second-largest continent in terms of both size and population (after Asia) (World Atlas, 2016). Africa's growing footprint between 1961 and 2008 largely arises from population increase, as there was a 5 per cent decline in the per capita EF in the same period (AfDB and WWF, 2012: 14). This decline runs counter to all other countries in the world in the same period.

Within these very general figures, footprint averages vary for African countries. In descending order, Mauritius, Libya, Mauritania, Botswana and South Africa have EFs above a fair share, while Egypt, Namibia, Chad, Mali, Gabon, Tunisia and Ghana hover just above and below the fair share. Countries with EFs less than half the fair share include Burundi, Zambia, Mozambique, Malawi, Democratic Republic of the Congo, Rwanda and Eritrea (AfDB and WWF, 2012: 12). Within these averages are many people with insufficient access to resources, who live well below the poverty line. As Irogbe (2013) asserts, in the more than 40 years since Africa was freed from colonial rule, "the region remains mired in famine and poverty". Before climate change became accepted by governments (or at least by most governments), drought was acknowledged to be Africa's chief natural disaster (Benson and Clay, 1998: 7).

Africa faces a projected population increase. The 1.02 billion people in 2010 are projected to reach 1.93–2.47 billion by the middle of the twenty-first century (AfDB and WWF, 2012: 35). In 2010, the people of Africa formed 14.8 per cent of the 6.9 billion people in the world (Population Reference Bureau, 2010). This will rise to around 22 per cent of a total world population of 9 billion by 2050. Birth rates of five children or more for each woman are still current in tropical African countries (Zinkina and Korotayev, 2014). This population increase will mean that Africa as a whole will have to draw on resources outside its boundaries as its footprint grows to be larger than its bio-capacity. This is the background to the stories.

Of the five African stories, the first is set in urban Morocco. Morocco has a rising population and a rising EF, moving from 0.96 gha in 1961 to 1.48 in 2010 (Galli, 2015). The second story comes from rural Mozambique, a country with an average EF in 2012 of 0.87 gha/person (GFN, 2016a), and the third from Johannesburg in South Africa, a country with an average EF of 3.3 gha/person in 2012 (GFN, 2016b). The fourth and fifth stories are both from South Sudan. In 2006, Sudan had an EF of 1.6 gha/person (AfDB and WWF, 2012: 12). The EF of South Sudan, which became an independent country in 2011, is not available. Civil war split the country and the peace is still tentative. The settings of each story are briefly described in the next sections.

Marrakesh, Morocco

Marrakesh lies inland in the fertile northern part of Morocco, a country with a long Atlantic and shorter Mediterranean coastline. Established in the ninth century CE, for many years the city was "a major political, economic and cultural centre of the western Muslim world" (UNESCO, 2017). Until the twentieth century, its power led to the whole country being known as the Kingdom of Marrakesh (Dorsey, 2015). Within Marrakesh, the 700-hectare Medina, where the story is set, remains a "living historic town with its tangle of lanes, its houses, souks, fondouks [places of business], artisanal activities and traditional trades" (UNESCO, 2017). The Medina is one of the reasons Marrakesh has long been a tourist city (Cunningham, 1992). Marrakesh has a semi-arid climate with hot summers and warm winters, with spring and autumn being hot but comfortable. Rainfall is low and sunshine hours high. Night temperatures are much lower than those of daytime (Holiday Weather, 2017).

Inhassoro, Mozambique

Inhassoro on the Indian Ocean is a small fishing port that has recently started to grow because of tourist initiatives. Mozambique was colonised by the Portuguese in the sixteenth century and became a centre for slave trading in the eighteenth and nineteenth centuries. It gained independence in 1975 (BBC News, 2017).

The country borders South Africa to the south, along the coast of the Indian Ocean. Inland, it borders Zimbabwe, Zambia and Malawi, and to the north, Tanzania. The Zambezi River runs through the centre of the country and Inhassoro lies to the south of the river. Since the end of colonisation, Mozambique has had internal conflict. Initially, it looked to the Soviet Union, which had provided support for the freedom fighters, and a socialist government, under the FRELIMO party, was set up after the Portuguese departed. However, the neighbouring countries of South Africa and the then Rhodesia wanted a Western-style regime and eventually sponsored the RENAMO guerrilla movement (Lamba, 2008). Although peace was signed in 1992, as the story below reveals, this conflict has never truly gone away.

Johannesburg, South Africa

Johannesburg, with its population of nearly 4 million, is South Africa's largest city. The city started life in 1886 as a gold settlement, and by 1919 was producing 40 per cent of the world's gold (Kracker Selzer and Heller, 2010). The city is now critical to the economy of South Africa, and arguably of all Africa south of the Sahara (Rogerson and Rogerson, 2015). Since the demise of apartheid, the former white supremacy has given way to a rising native South African middle class, within which the story is set.

Juba, South Sudan

South Sudan is the youngest country in Africa, becoming so in 2011 after a referendum (Thomas, 2015: xiii). The White Nile passes through the middle of the landlocked country and feeds significant freshwater wetlands. Juba, the capital city, sits on the White Nile. The country, which was once the site of slave trading, has oil resources, but continuing conflict has interrupted moves to prosperity and as the stories below reveal, has seriously affected the lives of its citizens. Juba has been the scene of fighting (Al Jazeera, 2017) and is home to Protection of Civilians (PoC) sites. "PoC sites refer to situations where civilians seek protection and refuge at existing United Nations bases when fighting starts. Although most UN peacekeeping missions have encountered this phenomenon at one stage or another, the creation of PoC sites on such a scale at the bases of the UN Mission in South Sudan (UNMISS) is arguably unprecedented in UN history"(Lilly, 2014).

Ismail and Mariam: Life in the Medina: Morocco

Researcher and writer: Fabricio Chicca

The house and contents

Ismail, 43 years old, lives in Marrakesh with his wife Mariam, 29 years, and their two boys Amine and Hakim, aged 10 and 11, respectively. They live inside the old Medina in a comfortable 82 square metre dwelling of plastered brick on the top of their shop. The house was entirely renewed in 2004 after a fire had damaged part of it. The traditional earth material of the original house was replaced by conventional bricks which support the concrete slabs that form both floors. The dwelling is narrow and long. On the ground floor, the family shop occupies the entire 4-metre width of the main façade. A spiral staircase located in the back of the shop gives access to the two residential floors. The majority of family life happens on the first floor, which is divided into a kitchen area for cooking, a toilet and a living room, where Amine and Hakim sleep. The living room has the traditional sofa attached to the walls with a small coffee table in the middle. On the second floor, there is a room for Ismail and Mariam, a bathroom and a small space used as storage for the shop, for personal belongings and the boys' clothes. There is a small outdoor area on the second floor, which Mariam uses to wash and dry clothes.

The kitchen has a fridge, stovetop and two ovens, microwaves and a small (30-centimetre) television. The living room, which is also the boys' room, has a larger (80-centimetre) television; the sofas; a coffee table, which is also the dining table; a broken air-conditioner; a ceiling fan; and a desk with a computer. The family have recently acquired an internet connection. Ismail and Mariam's bedroom has a ceiling fan and a wardrobe. The boys store their belongings partly in the space beside the couple's room and partly in a small chest of drawers beside the toilet. The chest of drawers is also used as a desk for their homework. The family's monthly average electricity consumption is 404 kilowatt hours, which includes the power for two refrigerators in the shop (a small 110-litre and a medium 240-litre refrigerator).

Transport

The family do not have a car as their entire social network is inside the Medina, with the boys' school located just outside it. Ismail's family have a long tradition, which he claims goes back 400 years, of running shops inside the Marrakesh Medina. The shop sells a wide variety of products, from snacks to make-up and medicine for both locals and tourists. The family's life happens around the shop, in the Medina and in their relationships with relatives and neighbours. The shop opens 6 days a week, except for Fridays when Ismail goes to the Mosque during the morning, after which the family visit relatives inside the Medina.

Routine

During the week, the family day starts at 6 am. While Mariam prepares breakfast, Ismail goes down to the shop to check the stock and receive deliveries of goods. He also goes out and buys anything they need for breakfast, which the family eat together at 7 am. During the week of this investigation, the family had *kobhz* bread (a circular wholemeal flat bread) with honey and olive oil, mint tea, fruit (oranges, figs and watermelon) and goat's cheese. After the meal, the boys go to their school which is a 20-minute walk through the Medina, with the school located just outside, near the north gate. Ismail goes down and opens the shop around 7:30 am, while Mariam prepares mint tea to be sold with bread to those passing by. After preparing the tea, she joins Ismail in the shop and they work together until 11 am, when she goes up to prepare lunch. Occasionally, one of them leaves the shop, as two people are only necessary when the demand is high (normally after 5 pm). However, most of the time, both Ismail and Mariam are either in the shop or sitting outside it socialising with fellow shoppers or family and friends.

During the week, on Tuesday and Friday, Ismail went to visit his parents, who live 5 minutes away inside the Medina. He also went to the Mosque with his father and cousins. Mariam went to her mother's house every day, usually in the afternoon when the shop has few customers. Her mother lives on the other side of the Medina, 20 minutes away, and Mariam and her sisters help her with her daily routine. The visits are very short and do not last more than half an hour.

The family's entire social life happens inside the Medina, and the family are well connected with neighbours and family. Often Ismail or Mariam sit outside their shop, or in front of someone else's shop nearby to drink tea. This socialising is very important to keep the relationships with the community. There are several informal gatherings during the week, normally based around tea drinking. Around midday, the boys come back from school for lunch. Depending on the number of customers or clients, the shop is either closed and the family gather around the tajine, or the tajine is eaten in the back of the shop. Ismail is responsible for deciding where to have lunch, which he does around 11:30 am every day, when he shouts to Mariam where the lunch should be served. This has become a local joke, and daily, the neighbours anxiously wait for his shout. During the week, the shop was only closed once at lunchtime. After lunch Mariam tidies the kitchen, the boys return to the school and Ismail goes back to the shop.

In the hot summer months, the shop is occasionally closed for a couple of hours around midday, as there are few customers. Because this investigation happened in winter, which is high season for tourists, the shop was open for long hours to serve the tourists staying in nearby hotels and hostels. The shop is supplied by a middleman, Tarek, Ismail's cousin, who is responsible for bringing the goods Ismail orders. Every Monday and Thursday, before the shop opens, Ismail goes to the edge of the Medina to meet Tarek. He gives Tarek the list of products, and receives those from the last order. He brings the products back on a trolley as the family do not have a car. Ismail also receives some sale goods early in the morning such as bread, milk, fresh cheese and fruit.

The afternoons are calm periods in that area of the Medina. This is perhaps the most important time for socialising. This happens naturally without a schedule and is just part of daily life. Relatives and parents come and go as do friends and fellow shoppers. Ismail and Mariam join in as long as someone minds the shop. This socialising is a natural part of living.

Food

Their diet is very simple, with small variations. The breakfasts never vary. Other meals are based around the traditional tajine, always followed by fruit and tea.

The tajines are made with vegetables and meat—either beef or chicken, except for Fridays, when the family have the traditional "couscous day", with the couscous dish made in the morning and eaten for lunch and dinner.

Leisure

When the boys return from school around 4 pm, they finish their homework and immediately head off into the Medina. They meet their friends to play football in the quieter alleys, or other games. At night, during the peak hours for shopping, the boys go to the Jemaa el-Fna (the big square near the centre of the Medina), where tourists and locals patronise the food stalls. Both Amine and Hakim work at a cousin's food stall, helping to attract patrons. They both want to run shops in the Medina when they get older. Around 9 pm, when the frequency of clients starts to drop, they come back home, where they join their parents watching television until around 10:30 pm. One of the aspects of the family's daily routine is the importance of the television. The main television in the living room is permanently on whenever someone is home; it is used to wake the boys up and also gives the signal to sleep when their parents turn it off. The family follow international soap operas and international football.

Thomas and Hyacinta: Charcoal burners: Mozambique

Researcher and writer: Fabricio Chicca

The house

This family is made up of Thomas (aged 41), Hyacinta (aged 31) and their eight children. The oldest child, a boy, is 14 and the youngest only 2. They live in the rural area of Inhassoro, approximately 27 kilometres from the town of the same name and 180 metres from the Indian Ocean. Their house was built in 2009, using materials donated by a Portuguese non-government organisation (NGO), on a 20 metre × 30 metre site. It is made of bricks with a roof of straw and palm leaves, and has metal windows and door frames (which are falling apart because of rust). It and another 200 houses built at the same time make up the entire rural area of Inhassoro, which stretches 50 kilometres from Vilanculo, a town on the coast, to Jofane, which is inland and just south of the Save River. The houses were built by an international cooperative which hired local workers to build the houses. The family could not provide further information about the programme, as, according to Hyacinta, there was no involvement from the community, except for signing a form to register themselves for the programme. The houses were built several years after this signature. The lack of additional information was confirmed by a neighbour, who added that most of the houses have been already demolished and the materials either sold or used in a different way. The family's house had been through some alterations, especially the roof, where the original plastic sheet has been replaced with locally available materials. The floor is a mix of compressed soil and cement. The house is divided into two equal spaces, each of 3 metres × 4 metres. The living space, where the entrance is located, is the place where the family spend most of their time, where the food is prepared on rainy days and where the children sleep. The living room has three chairs, a small table, a wood stove and shelves for the kitchen utensils. The other space is for the couple and is also used to store the four mattresses that form the bedding for the entire family. The couple's room has a small cabinet where the family store their best clothes that they wear to church on Sundays, and some small belongings. According to the family, none of the 200 houses that spread from Vilanculo to a point 10 kilometres north of Inhassoro has water, sewerage or electricity supplies. The family also said that some of the houses were never used because they were too hot to live in.

Work and routine

Thomas is a fisherman while Hyacinta is currently unemployed. Her last fixed job a couple of years ago was as a cleaner for an NGO in Vilanculo, and it lasted for 2 months. Currently, she has the responsibility of taking care of the small food gardens in the backyard and around their plot. Thomas has been a fisherman since he was a child, when he worked with his father and older brother. His boat was inherited from his father when he passed away some years ago. The children do not now go to school, although the two oldest sons did some years ago. The family's financial situation and the distance to schools have put the children out of reach of education. None of the local families were sending their children to school, despite the fact that education is free in Mozambique. Some schools are far away and transportation, school materials and uniforms have to be provided by the parents, making the process of sending children to school not feasible for poor families. The best these families can do is to select one child and send her/him to the school in the hope of this child being the future caretaker of the family.

The family at this time were not having their normal routine. The social, political and more recent environmental events (drought) have made it hard for families in the region to keep their former lifestyles. An armed conflict in the north of the province, mostly around the national road (E1), is one source of this difficulty. The area between the provinces of Manica and Sofala has been the scene of an armed conflict between government supporters and an armed group called RENAMO. Because of the conflict, the E1 is only used when a certain number of vehicles gather together at army roadblocks before proceeding along the road as a group. Therefore, a trip to Chimoio, which would normally take 6 hours, is now totally unpredictable and can take more than 48 hours.

The uncertainty of travel as a result of the armed conflict has affected the capacity of fishermen from the region to sell their fish. Before the conflict started, Thomas would go to sea for 1 or 2 days, or as long as the ice lasted. Often, he used to come back without any catch. When he caught some fish he would go to Inhassoro and sell his catch to a middleman, who would re-sell the fish in Chimoio. In those days, he could save a small amount of fish for his family's consumption, and a smaller amount to be bartered for other goods in the town. Any leftover fish would be saved for his and Hyacinta's extended families. Despite the bad days with poor catches, Thomas considered those times to be prosperous as it meant the family did not need to grow any food in their backyard as the fish were enough to provide their very basic family needs. Since the start of the conflict in 2015, the situation is deteriorating, and one result is that the middlemen have now totally disappeared. The consequence of this is that the bartering in Inhassoro has ceased as the fishermen are flooding the market by trying to exchange fish they cannot sell for other goods. The family also face another problem, this time environmental, which is the lack of rain. According to them, it has been one of the driest seasons in the region in many years. The water in the community well is at the lowest level ever registered and the use of water for agriculture has been reduced. Soon they expect to be totally forbidden to use water for their crops unless the rain falls in record levels. The small gardens at the back of the property and around their plot are already struggling and production of manioc, beans, corn and sweet potato is very low and not enough to feed the family.

The current family situation dictates their entire routine. For the past months, the family have been engaged in gathering wood and producing charcoal from it for sale. Thomas occasionally goes out fishing, but only when they have managed to have a good day selling wood and charcoal. Because fishing is unpredictable, if the family only engage in fishing, when they do not catch anything, they starve. For this reason, they would rather have a lower but steady income. In order to provide some family income, all members have an important role. The four oldest children, with their parents, leave their house around 4 am to walk to the bushes 3 kilometres away from their house to gather sticks. Two of the children left behind go to the well to fetch the water ration to which the family are entitled (3 litres per person plus 10 litres for watering their crops). The family's new water buckets have lids, avoiding critical spillages. At the bushes, the sticks suitable for making charcoal are separated. The remaining sticks are put together and carried directly to the boat, and loaded on. The children who were left behind go to the beach later to look after the wood piles. The wood separated for charcoal is burnt. While they are burning the fresh sticks, the charcoal from the day before is put into bags and loaded on to the boat on top of the sticks previously loaded. When the boat is fully loaded, or more commonly when they run out of charcoal or wood, Thomas and his two oldest children sail south to sell it to the communities on the beach. The biggest buyers are the small hotels and hostels around Vilanculos and Bazaruto. As the tidal regime in the area is very strong, they normally sail south, then beaching near Vilanculos, sell the production during low tide. They then wait for high tide, when the boat refloats. If they still have some charcoal or sticks left they sail south again to try to sell these, returning home much later. Otherwise, they sell and return home directly. Each bag of charcoal (60 litres) is sold for Mozambican metical (MT) 40 (US$0.50), while the sticks are sold for MT 10 (US$0.10) for as many as the buyer can carry. On a very good day, they would make around MT 150 (US$2.00). Thomas is glad when, instead of money, the buyers barter for sticks with goods. This happens often, and in this week they bartered for corn, manioc, rice and toilet paper. While Thomas and the children are selling the production, the rest of the family keep producing charcoal or gathering sticks. Occasionally, before they head back home, they go to the village and buy groceries. However, most of the time, usually they wait and go to Inhassoro because prices there are lower. During this period, they did not visit the grocery store and all money was saved. The family managed to make MT 520 (US$7.00) during the week.

Food

In the mornings, the family share some manioc or sweet potato. The young children have a small amount of the same food to eat during the day, while the parents and older ones only eat again at night.

At night, they would normally have a similar diet of manioc with fried onions, but one night they had a fish (bartered for half a bag of charcoal), another night they had sweet potato with a cabbage and another night four eggs were shared by all of them with corn and beans. Despite the hard times, they do not consider themselves unprivileged as they have food every day, regardless of the fact that some of them go part of the day without food.

Leisure

The only social activity the family had in this period was to get together with their two closest neighbours. During these visits, they sit under a tree and chat for a couple of hours about the events of the day. The traditional visit to church no longer happens as the family have to engage in their work daily, including Sundays. They hope one day to return to their old lifestyle. In those days, Thomas would not go fishing on Sundays, and the family would go to the church, which was about a 2-hour walk from their home. There, they would spend a couple of hours after the worship with the community. They would also visit Inhassoro fortnightly for groceries.

Dora and Arthur: A rising middle-class family in Johannesburg: South Africa

Researcher and writer: Dashakti Reddy

Background

Dora is 38 years old and lives in Johannesburg with her 41-year-old husband, Arthur, and their three children, aged 15, 6 years and 2 months. The couple were married in 2008. Dora was born and raised in Swaziland, along with her two sisters. She attended a Catholic primary and high school and went on to study contemporary music at university. The family have pets and speak English at home. Dora works in the financial sector as a team leader for a large bank and Arthur works in the private sector. Dora recently had a baby and is currently on a 4-month maternity leave. The family recently employed a home cleaner. They consider themselves to be middle class.

Housing

The house, of 80 square metres, is constructed of brick and concrete and is on the upper floor of a duplex setting, typical for South Africa as it offers an increased level of security. There are three bedrooms, a bathroom with a flush toilet, a living room area and a kitchen. Piped and treated water is supplied by the city council and is used for all drinking, cooking and bathing purposes.

Daily routine

The typical daily routine for the family includes Arthur leaving the house around 4:30/5 am to go for a 1-hour run. Dora's daughter wakes around 5:30 am to shower. Once Arthur returns, around 6 am, he wakes their son and bathes him while Dora prepares breakfast and lunch for the children. While Arthur is getting dressed for work, Dora dresses her son and then he watches cartoons and eats breakfast with his sister. Arthur's work place is right outside their house, so he walks there. However, first has to drive the children to school, which takes between 15 and 30 minutes, depending on traffic. Once he drops the children, he leaves the car at home and then walks to work. At the moment, as Dora is on maternity leave, this is the current schedule, but typically she would drop the children off at school and then drive to her work place and Arthur would walk to work.

Arthur comes home for lunch every day for around 30 minutes between 12:30–1 pm. Usually Dora would make his lunch in the morning before going to work but now that she is home, she often makes his lunch at lunchtime. He finishes work at 4:30 pm, picks up the children from school and arrives home around 5:30 pm, depending on the traffic. Dora's 15-year-old daughter finishes school at 2:30 pm, but stays on for extracurricular activities including cricket, guitar and piano. The days she does not have sport/music, she stays in the school library and completes her homework and then gets picked up around 5 pm.

Now that Dora is home, she prepares dinner during the day, around 2 pm, so it is ready when the family get home. When she is at work, she would generally start making dinner when she gets home around 5 pm. The family eat dinner together around 6–6.30 pm, and Dora starts winding her son down for bedtime around 7:30 pm. He has a bath and is in bed at 8 pm. Until his bedtime, Dora's daughter "doesn't like watching TV, because it's cartoons 24 hours", as these are the programmes that are left on to please her 6-year-old brother. When both children are in bed, Dora and her husband watch TV, but Dora noted that she also tries to sleep when the children and baby are sleeping.

Saturdays are typically for "lazying around and tidying up". More recently, the family have not been out much since the baby was born, but Dora explains that, "[on] our typical weekend, we just love being home ... we don't have a set time we wake up, but obviously my husband wakes up and [goes] to the gym and [for a] run. My daughter can sleep the whole day, my son wakes up early".

Sundays are very busy days for the family, as they are all heavily involved in their church: "we're all involved, my daughter plays the piano, I'm in the choir and my husband directs the TV department". Typically, on a Sunday, she explains that the "earliest we wake up is at 5:00am, my daughter is also a leader in the youth [group], if she's not scheduled in the instruments, she's in the youth". Church is only 2 minutes away, but the family drive because of the children. They leave home at 7 am, with the first service starting at 8 am. At 9:30 am, the church provides something to eat and at 10:30 am, the second service starts. The family are back home around 12:30/1 pm and eat a small meal, followed by a nap. Arthur also "does the evening service around 18:00, leaving home around 17:00 just to prepare, [and] at 20:30/21:00 the kids are already in bed to be ready for school the next day".

Food and eating habits

Dora and her husband have recently made an effort to change the family's eating habits. "We don't go out much to eat out because my son has gained a lot of weight from all that, that's why we're changed the way we eat. We're now on a mission, we're just having our greens and meat and veggies. I can't remember the last time we went out, we try to eat all the good things". The family have reduced snack eating—they still eat popcorn, biscuits and tea, but no sweets and chocolate. The family rarely eat out, maybe "once or twice a month, if that". When they do go out to eat, they typically eat "salads and your meats, typically we eat pizza, steak". The family's most common eating habits include eating an "early supper and maybe a coffee after that or tea with a biscuit".

During the week, the family eat cereal for breakfast, which is either cornflakes or Weet-bix. Dora's son is provided with lunch at school, but she usually packs him a snack consisting of yoghurt, a juice and a fruit for break times. Dora's daughter eats her packed lunch which includes a sandwich, typically cheese, chicken mayo, polony (meat) or egg along with fruit, yoghurt and juice. Arthur goes home for lunch and eats leftovers or a sandwich which Dora has prepared before going to work. When Dora is at work, she packs her lunch and eats it there but now that she is at home and breastfeeding, she has no set eating times. In the morning, she drinks tea and then snacks on biscuits throughout the day. For lunch, she eats a sandwich or leftovers from dinner. For dinner, the family typically eat one type of meat, usually chicken, with vegetables. On Sunday, they tend to eat something light like sandwiches. Dora explained that she rarely cooks a proper meal for lunch or dinner on Sunday due to time pressures. During the discussion for this chapter, the only changes to the family's eating schedule included having a family meal with Dora's parents on Saturday night and lunch at Dora's sister's house on Sunday (shown in italics in Table 2.1).

TABLE 2.1 The family's meals for a week in April 2016

	Breakfast	*Lunch*	*Dinner*
Monday	Cereal (cornflakes/ Weet-bix)	Sandwiches, fruit, yoghurt, juice	Chicken, baked sweet potato, steamed butternut, steamed spinach, cabbage
Tuesday	Cereal (cornflakes/ Weet-bix)	Sandwiches, fruit, yoghurt, juice	Wors sausage, boiled potatoes, cabbage, chutney
Wednesday	Cereal (cornflakes/ Weet-bix)	Sandwiches, fruit, yoghurt, juice	Chicken, baked potato, green beans, tomato chutney
Thursday	Cereal (cornflakes/ Weet-bix)	Sandwiches, fruit, yoghurt, juice	Chicken, spinach, butternut, carrots, beans
Friday	Cereal (cornflakes/ Weet-bix)	Sandwiches, fruit, yoghurt, juice	Chicken breast, sweet potato, spinach, mushroom sauce
Saturday	Cereal (cornflakes/ Weet-bix)	Sandwiches, fruit, yoghurt, juice	Chicken, rice, butternut, spinach, roasted sweet potato *(dinner with parents)*
Sunday	9.30 am snack at church	*Lunch at sister's house*	Sandwiches (egg/cheese)

Transport

The family own one vehicle. This is a five-seater small petrol SUV. At present, the vehicle is used every morning to drop the children at school, and then again in the evening to pick them up. On the weekends, the family spend most of their time at home and only typically use the car to go to church or grocery shopping. The travel times in the evenings are generally longer, lasting approximately 1 hour compared with around 30 minutes in the morning due to traffic. The two older children have bicycles for recreational use but rarely use them. The distance from the family's home to church is 1 kilometre, and to Dora's work is 6 kilometres.

Energy use

The family's monthly consumption of electricity for April was Rand 850.00 (US $57.00), and electricity is provided by the city council. Due to the power crisis in the country, at some times the family experience "load shedding", and so can be without power for a period of a few hours up to a whole day, as Dora explained: "during the winter last year it was [as] often as twice a month. There was a time we had no electricity for a whole day. It was very inconvenient as some food and meat went bad in the fridge and had to be thrown away". Electricity costs rise in winter. The family do not own an air conditioner or heater. Electricity is used for lighting, the TV and kitchen appliances including the microwave, fridge, stove, kettle and the washing machine, which is run approximately twice a week. They do not own a clothes dryer but utilise outside drying lines, like most families in South Africa. The family do not have internet access at home, again typical of most South African homes.

Patrick: "We ran for ten years": South Sudan

Researcher and writer: Dashakti Reddy

Background

Patrick is a South Sudanese man born in 1969, in a village near the Ugandan border where his family home still stands. He has two wives and now lives in the capital city Juba with his second wife and their children. Patrick's first wife lives in the village with their children.

Patrick had 13 siblings in total, out of which he is the eighth born. His mother gave birth to four children in South Sudan before seeking refuge in Uganda due to the civil conflict between Sudan and South Sudan (Anyama War 1956). She then gave birth to six more children, including Patrick, in Uganda while in the refugee camp. On return to South Sudan, she had three more children, two of whom died. Out of her 13 children, six died under the age of 12 years.

Patrick began school when he was 9, walking a 28-kilometre round trip, 6 days a week. When war erupted in Uganda in 1982 and refugees from across the border came into South Sudan, he attended a school set up for refugees, which reduced his daily walk to 20 kilometres. At age 17, in 1986, Patrick's family went into hiding in the bush, which ended his schooling. In his own words, "We ran for 10 years from 1987 to 1993. We were in the bush and then in '94 we left to Uganda as a refugee".

In 1990, aged 21, Patrick married his first wife in the village. In 1994, Patrick and his family went to Uganda as refugees once again, and between then and 1997 they lived in the transit refugee camp. When he was 29, he went back to school in the camp for 3 years. In 2002, he left his family at their settlement in the bush and went to Kampala for a driving course. However, in 2003, Joseph Kony's "Lord's Resistance Army" chased Patrick's family and the other people from the area out of the bush and displaced them to the town of Adjumani. Patrick's family remained there till 2008, although in 2004 Patrick decided to return to South Sudan to find work. He lived and worked in Juba and sent money back to his family. In 2010, aged 41, he married his second wife and they built a house in Juba.

Housing

Patrick's cousin in Juba gave him land on a temporary basis to build a house on. However, once Patrick gets his own land, his family will move. He has built a *tukul* (a cone-shaped mud hut) for his family and another small one for his older, unmarried brother in the same compound on his brother's land. Patrick's cousin does not allow him to construct any additional rooms as he says then Patrick will not want to leave. Nine people live in the house in Juba, four adults and five children. Patrick does, however, own his other family's land in their village. The house in the village is home to seven people, two adults and five children.

Patrick purchased the materials used to build his families' houses in both places—poles, mud, small amounts of cement and iron sheets for the roof. The house in Juba is 4 metres × 4 metres in a compound that is 15 metres × 20 metres, and is fenced with bamboo. The house consists of two rooms, one for Patrick, his wife and their new baby, and the other for the children. The children's room has two big beds, with the girls in one and the boys in the other. When Patrick's eldest son is home from boarding school, he has one bed to himself, with the others sharing.

Cooking is done in another *tukul*, the walls and roof of this one made of three sheets of iron. It is divided into two spaces, one for keeping food and one for cooking. The family own two 250-litre drums for storing water, which are located within the compound. These drums require refilling every 2 days by tanker at a cost of 30 South Sudan Pounds (SSP) (US$0.42) per drum. The water is used for cooking, bathing and drinking. The bathing room is located inside the compound and is again constructed from iron sheets. A bucket is filled with water from one of the drums and taken to this room for bathing. There is one pit latrine shared between all the inhabitants of the compound which Patrick dug himself. It is nearly full and he needs to build a new one but that will require buying six poles and three iron sheets (100 SSP per pole [US$1.4][1] and 200 SSP per sheet [US$2.8]).

Daily routine

Patrick is the head driver for an international NGO (INGO), and he has worked for it since 2006. He wakes at 6 am each day to help get the children ready for school. His wife takes them there while he minds the baby, and when she returns, he washes, has breakfast and goes to work. His breakfast consists of porridge made from millet paste with tea to drink. He leaves on his bicycle about 8:30/8:45 am and reaches the office around 9:30 am, sometimes at 9 am. At work during the day, Patrick does not eat lunch and only takes tea. He leaves the office between 6 and 6:30 pm, but sometimes stays until 8 pm if the office is busy. Patrick rides his bicycle home, takes a bath, eats and watches a DVD with his children. From around 9 pm to midnight the family sleep on mats outside because it is hot (40–45 degrees around this time of year), but at midnight they go inside with the window open because they are worried about thieves. Patrick works 6 days a week, and on Sundays he relaxes at home or goes to a plot he has nearby.

Patrick's first wife in the village has a job buying fish from Nimule and selling it in the village. For this, she has to travel approximately 30 kilometres using public transport. His second wife in Juba has no paid work. She prepares the daily meals for the family, cares for the children and the baby and goes to church on Sundays. Patrick's elder brother also lives with the family. He works as a lab technician but is a drinker and does not contribute to any household expenses. In South Sudanese culture it is expected that if someone has money or a job, they are responsible for caring for the extended family, no matter what the circumstances.

Food and eating habits

Patrick and his wife plan the family's meals at the beginning of each week, which is unusual for South Sudanese families. They write their decisions on a board in the compound so the whole family are aware of the plan, and can shop accordingly. Patrick buys the staple foods (millet, beans, onions, oil, tomatoes and groundnuts), charcoal for cooking and soap for the month, which are stored in the compound. Consequently, Patrick's wife only needs to walk to the market, located near the family's home, to buy the additional fresh items like meat and greens.

The family eat three times a day, but different members eat different meals. Patrick, his wife and his brother eat porridge and tea for breakfast every morning. Patrick does not eat lunch, to keep costs down, but drinks tea or water at work, both of which are provided. The children take tea in the morning before going to school, which is typical for most South Sudanese families. Once the children are at school, their mother prepares lunch. The children return home for lunch each day and eat with their mother. Lunch in South Sudan is between 1 and 2 pm every day, and most people return home for the meal. Patrick's brother eats lunch once he returns from work, also around 2 pm. The remaining food from lunch is then stored and eaten at dinner time. The family always eat together and never eat outside the home.

Patrick's family in Juba have no pets, but in the compound they keep two or three chickens which will eventually be eaten. In the village, they have two goats. The offspring of these goats will also be eaten (Table 2.2).

Transport

The family do not own any motor vehicles. Patrick owns a bicycle which he uses Monday to Saturday to travel to and from work. Cycling is unusual in Juba City and can be quite dangerous because of the traffic and motorbikes. The distance he travels each day to and from work is 30 kilometres. Patrick's brother walks to work, which is about 4 kilometres one way. The children also walk to their schools, which range between 300 metres and 1 kilometre from the home. Patrick's wife is the only person to use public transport, to get to church on Sunday, and this costs her around 10 SSP return, as the church is not too far from their home and on a main transport route.

TABLE 2.2 Family meals for a week in April 2016

	Breakfast	*Lunch*	*Dinner*
Monday	Millet porridge, tea	Beans with rice	Beans with rice
Tuesday	Millet porridge, tea	Greens with potatoes	Greens with potatoes
Wednesday	Millet porridge, tea	Beans with rice	Beans with rice
Thursday	Millet porridge, tea	Fish with potatoes	Fish with potatoes
Friday	Millet porridge, tea	*Dodo* (greens), cabbage, rice/*posho*★	*Dodo* (greens), cabbage, rice/*posho*
Saturday	Millet porridge, tea	Meat with potatoes	Meat with potatoes
Sunday	Millet porridge, tea Children take only tea	Beans with rice	Beans with rice

★Boiled maize flour

Energy use

The family's energy use is very minimal. They do not have mains electricity in their home. Patrick bought a generator in 2010 that lasted about 3 years before breaking down. He has not had the money to repair or replace it. South Sudan does not have any central power source provided by the government. All electricity is sourced from personal generators. In the last few years, there have increasingly been issues with this due to the scarcity and price of fuel and the consequences of war, despite South Sudan holding significant oil reserves. Patrick also has a small solar battery pack which cost him 1,000 SSP (US$14.1) for the battery and another 1,000 SSP for the solar panel. The solar pack is used to power a light bulb inside the house till about 9 or 10 pm and a DVD player which the family use every night to watch movies, usually from a flash disk. The family have not been able to use the television since the generator broke down, which is why Patrick purchased the portable DVD player. He also owns a small battery-operated radio and two basic Nokia phones for him and his wife. The family own no other appliances.

Grace: "I have to think what food these children will eat today": South Sudan

Researcher and writer: Dashakti Reddy

Background

Grace, who comes from Jonglei State in South Sudan, is 27 years old and the eldest of four children. She now lives in the UNMISS Protection of Civilians (PoC) site in Juba, the capital of South Sudan. In 1992, Grace's father took the family (Grace, her mother, two brothers and her sister) across the border into Ethiopia to seek refuge from the civil war between Sudan and South Sudan. Grace lived in a refugee camp for 10 years, where she attended the equivalent of 8 years of schooling. Although the schooling was supposed to be in English, the teachers and students mostly spoke Amharic, which she can now speak as a result. In 2003, Grace was married to a 23-year-old man. After the marriage, the couple travelled to live in the husband's village, where Grace continued her schooling. In 2004, at 16 years old, Grace gave birth there to twin girls.

Grace's husband was a soldier and he was deployed to Juba in 2007. She came to live with him there and continued her schooling, leaving the children as necessary with two other boys who were staying in their house. She eventually attained a diploma in financial accounting in Western Equatoria, where she gave birth to a boy in 2008. This was followed by another boy in 2011, born in Juba, and a girl, born in the PoC site in 2015. Grace moved back to Juba to live in her husband's house until the 2013 crisis, when civil war broke out there on 15 December 2013. She has been living in the PoC site since. In 2015, Grace's husband was shot and killed during the fighting and she is now the head of her household, an increasing trend for South Sudanese women.

In the village, her husband owned many cattle. He also had land outside Juba with a house that was taken over in the crisis. He had a big house in Bentiu but that was destroyed with its contents. On her mother's side, Grace said she also had many items, but everything was lost due to the crisis.

Current housing

Since 2013, Grace has lived on a small piece of land that was provided to those seeking refuge in the UNMISS base due to the civil conflict. The current shelter (Figure 2.1) is built with materials provided by the UN and INGOs. The materials include tarp sheets, wooden poles and nails. The shelter is approximately 2.5 metres by 4 metres, and consists of two very small rooms. It is shared by ten people—Grace, her mother-in-law, her husband's female relative, her three youngest children (the twins are in Ethiopia with her sister), two girls (her cousins, who her mother was caring for but who came to Grace when her mother died) and two boys (brother-in-law and male cousin). Space is a problem throughout the camp but also within the shelters. The boys sleep in one room, where food items are stored, and the girls in the other. Cooking is done on the ground just outside the shelter door (Figure 2.2).

Bathing areas and pit latrines were constructed and are serviced by the humanitarian community working within the PoC. These facilities are located within walking distance of the family's shelter and are shared by two blocks. For bathing, each person first collects water in a bucket from the water point, which is also not far from the family's shelter. The family have access to 20 latrines and 20 showers,

FIGURE 2.1 Typical shelters, PoC3, UNMISS, Author, June 2016.

FIGURE 2.2 Typical cooking facilities, PoC3 UNMISS, Author, June 2016.

evenly split between female and male. Grace explained that the population in the block she uses is not too bad but that larger populations in other blocks face challenges with these shared facilities.

Daily routine

Grace's daily routine starts with her thinking about what the family could eat, so it would be different from the day before. Her major problem is that the food given by the UN is not enough for the family of ten. She has to get money and buy additional food, including milk for the baby and charcoal for cooking. Grace describes her morning routine.

> My routine in the morning, I'm the first person to get up, and I'll put fire, prepare porridge for the children and put in a jug, and also prepare milk for the younger one, and then now from there I go to bath and brush my teeth and finish everything there and then I come to dress. If I have something (money) in my hand I'll give it to my mother-in-law, because all the girls are in school so the money is left at home, for them to go and buy food, because I've already instructed them to either buy fish or meat or greens. Then I come to work.

Between 1 and 2 pm, Grace goes home for lunch, which has been prepared by her mother-in-law. Before returning to work, Grace makes sure that there is something for the evening meal. When work finishes at around 5 pm (often sooner), she first eats and after supper, the family talk until it is time for bed, when everyone goes inside to sleep. On Saturday and Sunday, she does the family laundry and also all the cooking. She also goes to the market. Grace's mother-in-law's daily routine consists mostly of caring for Grace's youngest child. The children's daily routine revolves around attending school in the camp.

Even before her husband died and when he joined the army, Grace supported the family, her sister in-law and her children and mother in-law. She sent her twins to Ethiopia to her sister but still supports them with food and school fees. All this is a struggle for her.

Food and eating habits

The family are provided with 50 kilograms of maize or sorghum per month by the humanitarian community. Grace prepares the porridge in the mornings for the children. During the week, her mother-in-law prepares lunch for the whole family. The girls prepare the evening meal once they return from school. The family eat all meals together and in the home, returning for lunch each day because there is nowhere else to go and eat, and they can only afford to eat at home. The family do not have tea because Grace cannot afford sugar, a staple in South Sudanese tea culture. An additional stress and cost comes from Grace's mother in-law's eating habits. She only eats traditional *walwal* (sorghum mixed with water) and milk, which means having to buy milk every day. Table 2.3 sets out the rest of the family's meals for a week.

Within the PoC, the majority of food (including the maize, sorghum, beans and lentils) is distributed by the humanitarian community. Grace has a job, which allows her to purchase additional ingredients, such as meat and fish, which may not otherwise be available to the general internally displaced persons' community (Table 2.3).

Transport

The family do not own any vehicles but have access to public transport. Grace uses this on a Saturday and Sunday to travel outside the PoC, either to Jebel Market or

TABLE 2.3 The family's meals for a week in April 2016

	Breakfast	*Lunch*	*Dinner*
Monday	Porridge	Dried fish and sorghum★	Beans and sorghum
Tuesday	Porridge	Beans and sorghum	Dried meat and sorghum
Wednesday	Porridge	Greens and sorghum	Fish and sorghum
Thursday	Porridge	*Adase* (lentils) and sorghum	Beans and rice
Friday	Porridge	Meat and sorghum	Fresh fish
Saturday	Porridge	Fish and sorghum	Beans and sorghum
Sunday	Porridge	*Adase* (lentils) and sorghum	Greens and sorghum

★Staple can be maize or sorghum, depending on the monthly distribution

Custom Market. To reach either, she has first to pay for a ride on a motorbike and then a car to Custom Market, which costs 32 SSP (US$0.45). If there are no cars, the roundtrip on the motorbike costs 100 SSP (US$1.4). When there is no fuel, which is frequent in the war-torn country, Grace walks 4 hours to Custom Market, which is a 32-kilometre round trip. To walk to Jebel Market is a 13-kilometre round trip.

Energy use

Like the rest of the country, the family do not have electricity in or outside the PoC. Solar-powered security lighting is generally provided by the UN or humanitarian sector, but often gets stolen. Electricity is only provided through personally owned, fuel-operated generators. The family use collected firewood for cooking or charcoal, when they have money to buy it, as it saves cooking time.

Note

1 Exchange rate for March 2017, https://www.mataf.net/en/currency/converter-USD-SSP?m2=100.

References

African Development Bank (AfDB) and World Wide Fund for Nature (WWF) (2012) *Africa Ecological Footprint Report*, available at http://www.footprintnetwork.org/content/images/article_uploads/africa_efr_english_low_res_1.pdf, accessed 8 March 2017.

Al Jazeera (2017) *South Sudan crisis: Renewed fighting in Juba*, available at http://www.aljazeera.com/news/2016/07/south-sudan-security-council-demands-ceasefire-160711043656662.html, accessed 13 March 2017.

BBC News (2017) *Mozambique profile – timeline*, available at http://www.bbc.com/news/world-africa-13890720, accessed 10 March 2017.

Benson C and Clay E (1998) *The Impact of Drought on Sub-Saharan African Economies*, Washington, DC: The World Bank.

Brantlinger P (1985) Victorians and Africans: The genealogy of the myth of the dark continent, *Critical Inquiry*, 12(1), pp. 166–203.

Cunningham, R (1992) Fostering community: A significant role for the Medina souq, *Arab Studies Quarterly*, 14)(1), pp. 61–75.

Dorsey, J (2015) The Marrakesh Medina, *World and I* 30 (12), available at go.galegroup.com/ps/i.do?p=AONE&sw=w&u=vuw&v=2.1&id=GALE%7CA438358602&it=r&asid=0220e539c708146120b64ecd352dc531, accessed 15 May 2017.

Galli A (2015) On the rationale and policy usefulness of Ecological Footprint Accounting: The case of Morocco, *Environmental Science and Policy*, 48, pp.201–224.

Global Footprint Network (GFN) (2016a) Data sheets: Mozambique, available at http://data.footprintnetwork.org/api/b1/data/144/all/BCpc,EFCpc, accessed 10 March 2017.

Global Footprint Network (GFN) (2016b) Data sheets: South Africa, available at http://data.footprintnetwork.org/api/b1/data/202/all/BCpc,EFCpc, accessed 10 March 2017.

Holiday Weather (2017) *Marrakesh*, available at http://www.holiday-weather.com/marrakesh/, accessed 15 May 2017.

Irogbe K (2013) The persistence of famine in Sub-Saharan Africa, *The Journal of Social, Political and Economic Studies*, 38(4), pp. 441–461.

Jones J (2010) *In Search of Brightest Africa: Reimaging the Dark Continent in American Culture, 1884–1936*, Athens, GA: University of Georgia Press.

Kracker Selzer A and Heller P (2010) The spatial dynamics of middle-class formation of postapartheid South Africa: Enclavization and fragmentation in Johannesburg, *Political Power and Social Theory*, 21, pp. 171–208.

Lamba I (2008) Conflict prevention and resolution in Africa and lessons from the past: The Democratic Republic of Congo up to 2003 with comparisons from Mozambique, *Law and Politics in Africa, Asia and Latin America*, 41(2), pp. 146–168.

Lilly D (2014) Protection of civilian sites: A new type of displacement settlement? available at http://odihpn.org/magazine/protection-of-civilians-sites-a-new-type-of-displacement-settlement/, accessed 3 March 2017.

Pimm S (2007) Africa: Still the "dark continent", *Conservation Biology*, 21(3), pp. 567–569.

Population Reference Bureau (2010) *World Population Data Sheet*, available at http://www.prb.org/pdf10/10wpds_eng.pdf, accessed 10 March 2017.

Rogerson C and Rogerson J (2015) Johannesburg 2030: The Economic Contours of a "Linking Global City", *American Behavorial Scientist*, 59(3), pp. 347–368.

Stanley H (1878/1890) *Through the Dark Continent*, London: Sampson Low, Marston, Searle, and Rivignton.

Thomas E (2015) *South Sudan: A Slow Liberation*, London: Zed Books.

UNESCO (2017) *Medina of Marrakesh*, available at http://whc.unesco.org/en/list/331, accessed 15 May 2017.

World Atlas (2016) Africa, available at http://www.worldatlas.com/webimage/countrys/af.htm, accessed 10 March 2017.

Zinkina J and Korotayev A (2014) Explosive population growth in tropical Africa: Crucial omission in development forecasts—Emerging risks and way out, *World Futures*, 70(2), pp. 120–139.

3
ASIA

Fabricio Chicca, Jacqueline McIntosh, Yukiko Kuboshima, Jestin Nordin, Tri Harso Karyono and Brenda Vale

Introduction

Brenda Vale

Asia is the largest continent in terms of size and population and hence has more stories than any other chapter in this book. Asia, including the countries of the Middle East, forms approximately 30 per cent of the land area of the world (Merriam-Webster, 2007), spreading from Turkey in the west to Indonesia and parts of northern Russia in the east. This means the climate is very diverse, from countries like the Maldives and parts of Indonesia, that lie on or very close to the equator, to northern Russia in the Arctic Circle. Further extremes come from Asia being the site of the highest mountains in the world, all found in the Himalaya and Karakoram ranges which lie to the south and west of the Tibetan plateau.

The continent contains 48 countries, with Russia and Turkey both lying partly in Asia and partly in Europe. The overall population of Asia is 4.4 billion, or just under 60 per cent of all the people on earth. It has a young (average age 30) population (Worldometers, 2017). Table 3.1 shows the sub-regions of Asia with their populations and where the stories are located.

Asia is a continent with a wide variety of large and diverse eco-systems that have to compete with the activities of its peoples (Asian Development Bank [ADB] and World Wide Fund for Nature [WWF], 2012: 5), especially as most countries are still in a process of development and urbanisation. In 2014, 48 per cent of all Asians lived in urban areas, compared with 54 per cent of all people globally in the same year (United Nations, 2015: 10, xxi). Asia is also home to the three countries considered to be 100 per cent urbanised—Singapore, Hong Kong and Macao. The projection is for Asian urbanisation to continue, which has been viewed as both the cause of and the solution to the problem of environmental degradation; the solution comes if Asian urbanisation moves away from the "brown" development path

TABLE 3.1 Asian sub-regions, population and story locations

Sub-region	Population (billions)	Story	Location	Story population (billions)
Southern Asia	1.9	India	Rural	1.3
Eastern Asia	1.6	Mongolia	Urban	0.003
		Japan	Urban	0.1
Southeast Asia	0.6	Myanmar	Rural	0.05
		Malaysia	Urban/suburban	0.03
		Indonesia	Rural	0.26
Western Asia	0.26	–		
Central Asia	0.07	–		

TABLE 3.2 Summary of EFs and urbanisation of story countries

Country	EF (GFN, 2016) gha	2015 % population urbanised (World Bank Group, 2016)	Average annual urbanisation rate 2000–2015 (CIA, n.d.a)	Average annual population growth rate (CIA, n.d.b) (2106 est.)
India	1.2	33	2.4%	1.2%
Indonesia	1.6	54	2.7%	0.9%
Japan	5.0	93	0.6%	−0.2%
Malaysia	3.7	46	2.7%	1.4%
Mongolia	6.1	72	2.8%	1.25%
Myanmar	1.4	34	2.5%	1.0%

of the west and develops with "green" technology to create an urban middle class that will tend to be pro-environment (Wan and Wang, 2014). At present, however, because of its huge size, Asia is home to around 63 per cent of all the people in the world who live in rural areas (United Nations, 2015: 14). This split between high levels of both urban and rural living is mirrored by a wide range in Asian environmental footprints (EFs).

In 2008, Asia was already in biological capacity deficit with an average EF of 1.6 gha/person, which was almost twice the available biocapacity of 0.9 gha/person (ADB and WWF, 2012: 26). Since then, EFs for some countries, like Malaysia and India, have risen with development. More recent Asian EFs, using Global Footprint Network data from 2012, range from very high (UAE 7.9 gha), high (Japan 5.0 gha), above fair earth share (Malaysia 3.7 gha) and below fair earth share (India 1.2 gha) (GFN, 2016). EFs for 2012 are given for all the countries from which stories are presented in Table 3.2, together with an indication of the extent of urbanisation in each country. Table 3.2 shows that urbanisation is happening faster than population growth in the selected countries in Asia, which is problematic for EF as urban EFs are normally higher than rural equivalents in the same country (Vale and Vale, 2013: 58–59).

Although there is some description of the locations in the stories, this section introduces the wider background to each location.

Khajuraho, central India

The small town of Khajuraho bears the same name as a group of nearby temples that are World Heritage listed (UNESCO, 2017). Built in the tenth to eleventh centuries, these temples, some of which are highly decorated, are the reason tourists come to this area of India, which lies around 620 kilometres southeast of New Delhi. It is the remote and relatively undeveloped location that has ensured the survival of the remaining temples over the years. The town of Khajuraho is small, with a population of just under 25,000 in the 2011 census (Census Population, 2015). Khajuraho Nagar Panchayat (town authority) supplies water and sewerage to the 4,591 houses over which it has administration (Census Population, 2015). A study made 25 years ago found that tourism had not disturbed the traditional way of life as visitors fly in for a very short time just to visit the temples, although the same tourists have brought more prosperity through the employment that tourism offers to villagers (Jain, 1990). This same traditional way of living is reflected in the story of the family who live there.

Unlike the dry west of Madhya Pradesh, the central and northern parts, where Khajuraho is located, are served by a moderate monsoon (July and August), and so the land is relatively fertile (Khajuraho Tourism, n.d.). The summers are hot and dry, reaching as high as 47°C, but after the monsoon, the winter and spring (November to March) are pleasant, with a low of around 10°C (World Weather Online, n.d.).

Kendeng Mountains, Banten Province, Java

Banten Province lies at the western end of the Island of Java, across the sea from south Sumatra. This strait contains the island of Krakatoa, Indonesia's most active volcano, which was immortalised in the 1968 film *Krakatoa East of Java* (when in fact it is to the west). The Kendeng Mountains rise to 1,900 metres and form part of the range of mountains that runs across Banten from northwest to southeast (Ota, 2006: 14). Within these mountains live the Baduy people, who are very special when it comes to impact on the environment as their religion is one "that prioritises the principle of helping one another, spreading love and instilling a sense of responsibility among [sic] tribe so that there are no theft, robbery, adultery, destruction of the environment and dispute among the people" (Hamidi, 2016). They practice shifting cultivation to preserve soil fertility, and as well as planting rice, plant a tree legume as seedlings between the rows of rice, the legume helping to maintain soil fertility (Aweto, 2013: 149–150). A further description of where the Baduy live is given in the story below, including climate data.

Nagoya, Japan

Japan is a highly populated, economically successful and historically rich island nation with a shrinking population. The population of Japan grew steadily throughout the twentieth century, reaching a peak of around 128 million people in 2008, after which it has begun an equally steady decline. This is due to a shrinking birth rate and an ageing population, with people living much longer (Hara, 2015: 11–13). This is a concern because of the effect it will have on the economy, which is currently recovering following the world economic slow-down (Statistics Bureau, 2016).

The current population of Japan is housed in a built environment that occupies 5 per cent of the available land, with 67 per cent of the country covered by forest and 12 per cent given to agriculture, including paddy fields (Statistics Bureau, 2016). The Chukyo Metropolitan area is centred on the city of Nagoya, and is the third largest in Japan after Tokyo and Osaka. The climate is mild but with rain all year round, heaviest in the summer months. The August average temperature is 28.1°C and that of the coldest month, January, is 4.4°C (Climate-data, n.d.).

Penang, Malaysia

Penang is a state on the northwest coast of Malaysia, partly on the coast and partly on Penang Island, which houses the state capital of George Town. The state has a population of nearly 2 million. The family in the story live in the north of Penang Island, which is the most densely populated area (National Higher Education Research Institute, 2010: 3). The area is important for the development of northern Malaysia. From the 1970s, its industrialisation, which started with the making of electrical and electronic components, has moved to higher-value computer parts and consumer communication products (National Higher Education Research Institute, 2010: 12). By 2000, nearly 80 per cent of the population of Penang was urbanised (Masron et al., 2012). Nevertheless, with its sandy beaches, Penang is both a centre for tourism and the home of traditional fishing villages, which were disrupted by the December 2004 Asian tsunami (Krishnaswamy et al., 2012).

Penang, like Malaysia as a whole, has an equatorial climate with annual temperatures ranging from 23–35°C, with the hottest months being February and March and the wettest months being September and October (World Weather and Climate Information, 2016a).

Ulaanbaatur, Mongolia

Ulaanbaatur is both the capital city of Mongolia and the coldest capital city in the world (Caldieron and Miller, 2013). The hottest month is July, when the maximum temperature is around 27°C and the minimum around 12°C. In January, the coldest month, temperatures range from −10°C to a very chilly −25°C (World Weather and Climate Information, 2016b). Ulaanbaatur has been the site of significant recent urban expansion, with its population rising from 600,000 in 1989 to 1.03 million in 2007, forming 39 per cent of the population of the whole country (Kamata et al., 2010: 1). Because, historically, Mongolia has been a nomadic, pastoral country, land is state-owned and for the common use of everyone. It is also a country with a low population density and, since 2002, with urbanisation, each Mongolian household is entitled to a plot of land for housing (Caldieron and Miller, 2013). In Ulaanbaatur, this is a plot of 700 square metres (Kamata et al., 2010: 16). The urbanisation process starts with putting up the traditional *ger* (yurt) on the plot, which is then fenced. Building a more permanent home follows later. However, this means that infrastructure is not always in place when settlement happens, so water has to be fetched and rubbish is not collected (Bolchover et al., 2016: 25). Coal is the fuel used to heat these urban settlers, leading to bad air pollution problems. Ulaanbaatur is not just a cold capital city but a capital city in transition.

Lake Inle, Myanmar

Lake Inle lies in central Myanmar in the southern Shan Plateau (to the south and east of Mandalay in the centre of the country). It is a place where recent population growth and tourism have been changing traditional ways of life (Htwe et al., 2015). Tourist attractions include the floating gardens, where residents grow crops like tomatoes, beans, garlic and flowers on floating mats of elephant grass cut from around the lake edges where they have accumulated soils. The mats are floated onto the lake, where they are anchored with bamboo poles. However, this practice only dates to the 1960s, after the military government was established (Sidle et al., 2007). The lake is also the source of hydroelectric power for southern Myanmar so there are concerns over loss of lake area and silting from gardening (Su and Jassby, 2000).

The climate is very mild, with the average maximum monthly temperature never rising above 31°C (March–April) or falling below 24°C (December). The months of December and January are cold at night, with average minimum monthly temperatures of 4–6°C. The rainy season is from May to October (Myanmar Travel Group, 2016).

Ramesh and Anika: A day in rural India

Researcher and writer: Fabricio Chicca

The Indian rural family live on the outskirts of Khajuraho, which is locally known as Old City, in Madhya Pradesh Province in central India. The daily routine of the family, from their eating to their hygiene habits, was observed for a week while living with them. This analysis is focused on the items that are more relevant in terms of environmental impact, such as diet and transportation (Vale and Vale, 2009), together with leisure activities and cultural ceremonies.

The house

The family of three consists of parents Ramesh and Anika and their son Deepak, and so is smaller than the 2015 Indian average of 4.8 persons per household (Esri, 2016). Ramesh and Anika consider themselves a high-class family in Khajuraho Old City. Their house, which is on a dead-end street, was built in 1987, and is made of brick and timber. It has a single room with no partitioning. The house is approximately 6 metres by 3 metres, and the adjacent private external area is approximately 6 metres by 8 metres. The house has no electricity supply and no piped water. Water comes from collected rain stored in a tank beside the house. Water from the communal well located a hundred metres away is used for cooking and drinking (without any treatment), while the rainwater is mostly used for cleaning and showering. This family forms part of the large number of rural families which obtain water from a communal well. The toilet is a pit latrine outside the house, located in a public area, something that is quite common.

Daily routine and food

Ramesh works in Khajuraho, and Anika in the community agriculture field which is 1.2 kilometres from the house. The family wakes up around 6 am during the cold months of winter and around 5 am during summer. Their first meal of the day is rice and chickpeas served with spices and sauces. Throughout the week of observation, their goat was milked twice. The milk was mostly used in cooking, with no drinking of milk observed. Anika leaves home around 7 am, followed by Deepak, who attends the school in the vicinity. Ramesh leaves the house around 8 am to walk the 1 kilometre to the town where he works as a barber. He returns home in the middle of the day and the family gather for the midday meal. The food for this is mostly cooked the evening before, and small changes are made for each meal to give some variety. The diet is exclusively vegetarian and with few variations. Rice, chickpeas, potatoes and spinach are the main components. In the afternoon, Anika and Deepak go to the field, while Ramesh returns to town until 3 pm, after which he joins the others working in the field, which they continue to do until dusk. The fields are cultivated to provide food for the family and some surplus. However, in periods of low production, they use all the crops that they grow for their subsistence. The last chickpea crop was very successful so they could afford some small treats such as fruit and new clothes. During the investigation, the entire chickpea crop was harvested and they were constantly and proudly showing the entire production. That harvest would provide food for 2 or 3 months for the family of three. They also grow onions, potatoes, cabbage, tomatoes and spinach. Through the week, they had all their meals together. Rice was the base of all meals, and occasionally fruits were incorporated into the diet.

Transport

All family activities are reached by walking. The distances travelled are essentially the distance between the house and work or the house and school (around 1.2 kilometres each way). During the week described here, the family went twice to Khajuraho as small gatherings, mostly for religious purposes, happen almost every night in the Old City. On the last night, a larger meal was prepared for a religious celebration there. For this celebration, families brought different foods to be shared.

On this occasion, Ramesh, Anika and Deepak did not eat the food they brought but feasted on the dishes from other families, presumably as a way of bringing variety into their diet. During the weekend, the only observable difference in the family's routine was that Deepak joined Anika in the fields early in the morning as the school was closed. The family allow themselves a shorter work day once a week, normally on Sundays, when in the afternoon they join other families from the Old City in an informal gathering, considered to be the most important social event of the village. The men gather on one side of the largest street for smoking and conversation, while the women stay on the other side talking and taking care of the children.

Energy

The types of energy used by the family in this study are the same as those used by an ancient society, because they use a wood fire for cooking and candles for light. Ramesh's work in town is often paid for in goods, and any money he earns, as well as that earned from selling the crops they grow, is also mostly spent on extra food.

Slamet: A life unchanged for generations: Indonesia

Researcher and writer: Tri Harso Karyono

Background

The Baduy, who prefer to be called Kanekes, are a vernacular community living in the forest area of the Kendeng Mountains in the province of Banten, on the island of Java, Indonesia. With a population of about 11,500, the Baduy live in their homeland of about 50 square kilometres of hilly forests, about 300–500 metres above sea level (University of Iowa, n.d.). The average temperature of this village is about 20°C, which is considered cool for an average tropical climate. The homeland of the Baduy is about 40 kilometres away from the nearest city of Rangkasbitung, the capital of Lebak Regency, and 120 kilometres from the Indonesian capital, Jakarta.

The word "Baduy" may come from Baduy Mountain or Baduy River in the north of this area (Joshua Project, 2017). This vernacular community belongs to the Sundanese ethnic group, speaking the Sundanese language. They believe in a particular faith, namely Sunda Wiwitan, which is considered to be a mixed faith between Animism and Hinduism. The Baduy are divided into two sub-groups: the Baduy Dalam (Inner Baduy), with a population of about 1,500, and the Baduy Luar (Outer Baduy), with a population of about 10,000.[1] The community members of the Inner Baduy have more restrictions on their behaviour than the Outer Baduy. The Outer Baduy are used as a form of protection of the Inner Baduy from the outside world. No foreigners are allowed to visit the Inner Baduy; however, they can have contact with the Outer Baduy.

With the help of a friend, I met a young man in the Outer Baduy area. With his bare feet, he had long hair with a traditional headband, and was wearing a white shirt and a dark skirt like a Scottish kilt and had a white cotton bag on his shoulder. This was Slamet, a member of the Inner Baduy community. We took seats at a small food stall. After ordering cups of coffee and tea, I asked Slamet about his family and the place where he lived. He replied that his family lives at the Kanekes' village in the District of Leuwidamar, which is a part of the Lebak Regency in the Province of Banten. He added that his house was about 1.5 hours by foot from the food stall where we were sitting. The story below is narrated based on my interview with Slamet.

Slamet, 34 years old, started his talk by introducing some mystical taboos that the Baduy must obey. The Baduy are forbidden to kill, steal, lie, commit adultery, get drunk or cut their hair. They are not allowed to use modern technology to cultivate the soil, or any means of transportation, including motorised and non-motorised vehicles, bicycles and even horses or other animals. They have to travel with their bare feet since using footwear is also restricted.

The house

Slamet and his 32-year-old wife have three sons of 10, 8 and 5 years old. This family lives in a small neighbourhood as part of the Inner Baduy community. The family lives in a typical rectangular Baduy house of 48 square metres, plus an open terrace of 6 square metres. The house is elevated about 30 centimetres above ground level on short stilts resting on stone foundations. A terrace is located at the front of the house to be used mainly for receiving guests, followed by a single door to enter the main room of 30 square metres. The main room functions as the place for general family activities such as sitting, eating and sleeping for the three children. There is a parents' bedroom of 6 square metres on the front side of the house. A kitchen with its stove, including food storage, of total 12 square metres, is located at the rear side of the house, next to the bedroom. The house has no pieces of furniture; everything is done on floors made of woven bamboo. For cooking, the family use a stove with firewood. All the cooking devices are made from clay, wood and bamboo. Any device which contains non-organic materials, such as metal and plastic, is prohibited from use. Since electricity is forbidden, no fossil fuel energy is being consumed. During the dark night, the house is lit by bamboo lanterns fuelled by coconut oil. The main structure of the house is mainly made of various species of wood and bamboo. The walls are made from woven bamboo, while its pitched roof is covered with palm leaves. Slamet said that the palm-leaf roof is normally replaced every 5 years. No metal materials, including nails, are allowed to be used in the buildings of Inner Baduy. All construction joints are connected and tied up with organic materials such as rattan, bamboo and wood.

Daily routine and food

In the very early morning, the family go to the nearby river to take a bath and to toilet. Slamet went to the male side, while his wife and the three sons went to the female side. His wife also brought some pieces of dirty clothing to be washed. The small river is only about 50 metres away from his house. People in the community have built public baths for men and women in different places on the riversides. Children tend to follow their mothers into the women's baths. Slamet said that the Inner Baduy people are not allowed to use soap, toothpaste or detergent. This is in order to prevent the river water from contamination by non-organic compounds. They have to make sure that the river water is clean enough for domestic water purposes, including for cooking and drinking. Instead of using bath soap, people take a bath with parts of *Honje* plants, such as its flower, leaves and branches. *Honje*, or *etlingera elatior* in Latin, is also known as red ginger or wild ginger, with some different names in different places and countries (National Tropical Botanical Garden, 2017). For a toothpaste, they normally used mashed charcoal. To wash the clothes they use *Lerak* (*Sapindus rarak*) fruits (Fern, 2014); the trees of this fruit grow very well along the riversides.

Back from the river, the family take breakfast together. Their foods consist of rice, salted fish and small pieces of boiled vegetables. When rice is running out, they normally take sweet potatoes. Except for the salted fish, all the foods come from Slamet's farmland. The farmland is about 1 hour away by foot from his house. The Baduy are prohibited from farming in the yards around their houses. Any kind of farming must be done in their farmlands, away from their houses.

After breakfast, Slamet, with his wife and three sons, goes to the farmland to cultivate their agricultural crops, such as sweet potatoes, long beans, green beans, bananas and rice. The wife also brings some food for the family lunch. In the farmland, there is a hut of about 30 square metres in which the lunch is normally taken. The hut is built properly like the house, using wood and bamboo for all parts of the building and palm leaves for the roof. The hut is not only used during lunch time, but sometimes it is used for the family to stay overnight or longer, particularly during the rainy season.

The Inner Baduy are forbidden from cultivating cash crops; however, they can sell or barter some forest products, such as honey and durian fruit, to people outside the Inner Baduy territory. It takes about an hour for Slamet and his family to reach their farmland on foot. This farmland is part of the community's homeland. Every single family of the Inner Baduy has been provided with half to one hectare of dry land by the community to be cultivated to support their family life.

The Inner Baduy are not allowed to cultivate paddy rice in the wetland or *sawah*. They have to maintain the original land and soil as they are, minimising the changes. Cutting and filling the soil or changing the dry land to wetland are prohibited. They are also prohibited from using metal in cultivating the land; instead, they have to use wooden sticks to make small holes to put seeds in the ground. However, using metal tools is allowed to clear up the bushes and grass.

Because the rice is planted in the dry paddy field and has no fertiliser, except natural compost from the waste plants, it can only be harvested once a year, in comparison with normal wetland cultivation, where paddy rice can be harvested three to four times a year. However, Slamet said that in a normal situation, most of the Baduy families could save enough from a year's harvest of their rice for a year of family consumption. The Baduy do not sell the rice they harvest, instead consuming it for the family's needs. In case the rice runs out before the next harvest, Slamet or his wife can go to the small market outside his village in the Outer Baduy area, which is about 2 hours away from his house by foot. It is common for the Inner Baduy to visit this market once a week to buy some daily needs, such as salt, sugar, salted fishes, *tempeh* (soya beans) and some kinds of vegetables which are not planted in their fields. For a living cost, Slamet said that in a week the family has to spend about a hundred thousand rupiahs (US$8). He said that the spending is mostly for his children's needs. He regretted that today most of the Baduy children have been contaminated with such modern foods as biscuits and sweets. If they did not spend on these kinds of luxuries, the family might save half of their current expenses.

In the farmland, Slamet, his wife and the boys start to work together, digging the soil with wooden sticks and bamboos, clearing up some old and rotten banana leaves and also collecting rotten branches for firewood. They pick some bananas and also some young sweet potato leaves and long beans to be cooked the following day. In the middle of the day, they stop working and prepare for their lunch in the hut. The meals typically consist of steamed rice, grilled salted fish and some cooked vegetables. For dinner, they would eat similar foods to lunch. Normally Slamet's family would have the same meals almost all the time, although on one or two days a week they might have some fresh fish caught in the river. On some occasions, they might only have rice and salt for their meals when there is nothing left. Within a week they might have *tempeh* twice and chicken eggs or chicken meat once. Along with the main meals, they might also have some additional foods, such as steamed sweet potatoes, bananas or other fruits collected from their farmland or forest.

Slamet has a few chickens at home, from which his family can get eggs to eat once a week. The chickens are not caged, but roost in the branches of trees at night; they find their food themselves in the surrounding yards. The Baduy are not allowed to keep large domestic animals with four legs, such as goats, cows, buffalo etc. Also, they are not allowed to farm fish. They can catch fish in the rivers or nearby ponds by line fishing or using a fishing net.

According to Slamet, the Baduy very rarely eat red meat. On a number of ceremonial occasions, there will be some meat meals, such as in the traditional ceremony, which is held once a year, in the fifth month according to the Baduy annual calendar. It is common for the Baduy that, a few days before the celebration, some men will hunt deer in the forest to provide meat meals for this celebration. It is prohibited for the Baduy to have wild pig meat for this traditional ceremony, although such meat may be served at family celebrations, such as having a newborn baby or a marriage party.

The Baduy are happy to be visited by people from outside their community, since going outside their village to visit by foot would take time and consume energy. When Slamet was younger, he and his friends would visit Jakarta at least once a year by foot. It took 3 days to get to the capital city 120 kilometres from his village. They had to take short-cut paths during the day and slept at night at the places they were. They brought pure honey they had been collecting from the forest and sold it to people in Jakarta who had previously visited the Baduy village. Today, with his three sons, Slamet said things had changed and he had never thought to walk again to Jakarta for the same reason.

Fumio and Hana: A Japanese family

Researchers and writers: Yukiko Kuboshima and Jacqueline McIntosh

Background

This section describes the daily life of a Japanese family that were studied through the use of a questionnaire, the review of a detailed diary kept by the family for one week and the analysis of documents such as floor plans and electricity and gas bills, followed up with a telephone interview to clarify any outstanding matters. The family consist of a couple, Fumio and Hana, in their thirties, and their three daughters aged 6, 3 and 1. They are a nuclear family, a household type now prevalent in Japan. Having three children, they are stemming the tide of the declining birth rate with an average total fertility rate of 1.5 in 2013, a major social issue in Japan. Fumio works for a private company and Hana stays home to look after their children, which is typical for a nuclear family in Japan.

Housing

The family lives in a city with a population of 300,000, next to Nagoya City in Central Japan. Their house is in a residential area approximately 10 kilometres from the city centre. They used to live in central Nagoya City, but moved to their current house 2 years ago, seeking a better environment for their children. While most young families prefer to purchase new homes, Fumio and Hana bought a house from the 1970s, a time when a number of houses were built in Japan. Empty houses have become a major social concern in Japan and are linked to safety issues in many neighbourhoods due to arson, vandalism and other crimes. One factor in the oversupply of housing has been the unpopularity of second-hand houses, which is shown by their low market share of around 15 per cent, compared with 70–90 per cent in Western countries.

The house is a two-storey detached dwelling of approximately 120 square metres, surrounded by a garden. It is somewhat unusual as the main structure is of reinforced concrete, whereas the majority of stand-alone houses in Japan are of timber. The family has a vegetable garden and some fruit trees as well as other trees. On the ground floor, there is a large open-plan kitchen and living-dining space, a Japanese-style room and a large study, as well as the entrance, toilet and a bathroom. The upper floor consists of three rooms and a balcony. Everyone sleeps on mats in the largest upstairs room in the traditional Japanese style. The two currently unused rooms will be children's rooms in the future.

Daily routine

Their morning starts early, when the youngest daughter wakes up, usually between 5 and 6 am, depending on the season. On weekdays, after breakfast, Fumio leaves home around 6:30 am to arrive at work by 7 am. He works at a factory that operates in shifts, which is different from the most common type of work, which starts at 8 or 9 am and finishes at 5 or 6 pm, but often results in much longer hours. While the shifts change regularly, on days when he starts at 7 am, he generally returns home between 4:30 pm and 6:00 pm.

Hana's morning is organised around getting her two elder daughters to school and kindergarten. The oldest leaves for school at 7:45 am and meets with other children at a prearranged time and place so they can walk to school in a group. The second oldest goes to kindergarten Monday, Tuesday, Thursday and Friday. This starts at 9 am and finishes at 2 pm. Before driving her daughter there, Hana usually finishes the housework such as washing clothes, cleaning the house and putting the rubbish out. After returning home, while the baby is sleeping, Hana has time to bake bread and cook. After lunch, she generally does more housework and sometimes buys shopping on the way to/from picking her second daughter up from kindergarten. When the weather is good (except on Thursday, when the children go home in a group by themselves), she walks with her younger children to pick up her eldest daughter from school.

After school/kindergarten, the children play in the house and garden. Indoors, they draw, make Origami and play house, and in the garden they ride bicycles or tricycles, play house with picked flowers and climb trees. The oldest daughter usually does her homework before dinner. After dinner at 5–5:30 pm, they take turns to bathe, brush their teeth and go to the toilet and are in bed by 7 pm. Hana often also goes to bed at this time but sometimes she will do paperwork or hobbies such as sewing in her space in the Japanese-style room till 8:30 pm. Fumio stays up till 11 pm in the study, doing paperwork, checking emails and playing games on the PC. The family normally only watch the news on TV for approximately 30 minutes a day. They obtain information through the internet and by chatting with other mothers at kindergarten.

Variable activities involve visiting a friend for lunch or being part of a parents' gathering at the kindergarten. The oldest daughter takes swimming lessons after school on Fridays and is accompanied by Hana and her other daughters. On Wednesdays, when there is no kindergarten, they have more free time and sometimes go shopping at a large mall 40 minutes' drive from home, spending up to 2.5 hours there.

At weekends, Fumio often goes to work, although the day and times vary. When not working, he plays with the children, helps with housework and makes furniture for a hobby, while Hana does the heavy housework, including drying the mattresses and duvets, and looks after the baby. As a family, they play cards and picnic in the large park near their house.

Transportation

The family own one car, which is mainly used by Hana for ferrying her second daughter to kindergarten 1 kilometre away, and for shopping and occasionally taking her oldest daughter to the swimming pool. Hana would like to reduce her car use, but she finds it too convenient with her small children. She has arrangements with other mothers at the kindergarten to take turns collecting and dropping off the children, which saves time, effort and petrol and which is particularly important when someone is sick.

Fumio travels 7.5 kilometres from home to work by motorcycle. He started to use a motorcycle for commuting after he changed his job 6 years ago, and was no longer able to use the train as the new workplace was far from the railway station. Fumio and Hana could not afford two cars, so he chose to commute by motorcycle, which also uses less petrol.

The family walk (or the children sometimes ride bikes) when they go to school or the park 300–500 metres from home. They use the car several times a year when they travel any distance for recreational activities or to visit their parents' houses (1 hour away by car). The longest recent car travel was 4.5 kilometres, for a snow-playing event organised by the kindergarten. The family rarely use public transport (train or bus) or travel by air because the children are small.

Food and eating habits

Hana usually cooks all meals, and the whole family has breakfast and dinner at home, but lunch varies. Fumio and his second daughter usually eat their home-made lunch at work and kindergarten, while the oldest daughter has lunch provided by the school and Hana and the youngest daughter have lunch at home. Their staple diet is rice, but lunch is sometimes home-made pizza, pasta and Hana's home-made bread. The largest number of dishes in a meal is for dinner, with typically eight to ten different items. There will be three to four items for breakfast. The ingredients include various vegetables, mushrooms, chicken, pork, fish and other seafood. They drink green tea. For the home-made lunch box, Hana includes rice or rice balls with two to four dishes (Figure 3.1). Most days the school lunch is rice with milk and three dishes, except for one day a week when bread is served instead. The children's snacks between meals are non-confectionary such as baked sweet potatoes, strawberries, rice balls and home-made bread.

Hana cares for her children's health, which she believes is closely related to food and eating habits. The children rarely eat fast food and Hana actively avoids food additives. She buys pesticide-free and GM-free vegetables and fruit, grown locally as much as possible. She avoids food grown in the western areas of Japan to avoid radioactivity. The children eat mostly what Hana makes but Fumio eats food sold in convenience stores in addition to his home-made lunch. He also participates in his company's celebration drinking parties once a month on average. The rest of the family rarely eat out.

Hana shops at the supermarket but grows some food at home. The fruit trees include an *Ume* (Japanese plum) tree which is 35 years old and bears fruit in April. Hana grows peas, lettuces, tomatoes and eggplants. She serves the vegetables in the meals, and makes *Ume* juice and *Ume-boshi* (sour dried plums) once a year.

She composts waste food and the compost is used in the garden. She likes having a vegetable garden as the vegetables are fresh and pesticide-free and because her parents have been doing the same things since she was a child. When they all go to Hana's parents' house every 2–3 months, they are given vegetables such as sweet potatoes grown in the parents' field. When they visited Fumio's family to celebrate their daughter's birthday, they were given cakes and strawberries. Hana and her friends give each other food in exchange for dropping off and picking up children from kindergarten. They give dishes when a parent is sick and provide rice balls or snacks for the children.

Rice is steamed in a rice cooker once a day. The dishes are often prepared so they can be served in more than one meal. Leftovers are wrapped and kept in the fridge, then warmed up in the microwave oven. This provides the family with many meal options as well as saving time. Leftovers are more commonly used for breakfast than for dinner.

A typical dinner (Figure 3.1) has nine dishes with rice, and miso soup using home-made miso. There is boiled and seasoned food using bamboo shoots grown on a friend's mountain and peas picked from the garden; boiled and seasoned taro using taro from Hana's parents' house; stir-fry with pork and vegetables; a friend's home-made stir-fry with carrots and tuna flavoured with home-made salted rice malt; marinade of octopus and commercially bought *Natto*, which is fermented beans; and commercially bought *Mekabu* seaweed. Of these, four dishes had been cooked on previous days.

Consumer goods

The family try to reduce their consumption. For example, they use fabric nappies for the baby instead of paper ones and Hana and the children rarely buy new clothes. The children typically wear second-hand clothes given by families at the kindergarten and they only buy new clothes when the old ones are torn. They have fabric shopping bags to avoid using plastic bags. This is partly due to charges for plastic bags, a practice which is increasingly common in Japanese supermarkets.

FIGURE 3.1 Typical dinner (left) and lunch box (right).

Their electrical appliances include a TV, a washing machine, two vacuum cleaners, an iron, a hairdryer, a desktop PC, a telephone and a mattress dryer. This is used for drying the futon on which they sleep; traditionally it would be hung out in the sunshine. They do not often buy new appliances. They had to replace the TV 6 years ago when analogue TV broadcasting ended. They replaced their former car with a second-hand one 3 years ago, after using the old one for 9–10 years.

Energy

The climate is characterised by high temperature and humidity in summer and low temperature and humidity in winter. The average highest daily temperature in August is 33°C and the average lowest in January is 1°C. The time of year described here was from the end of April to the beginning of May, when the average temperature was 2–3°C higher than the annual average.

The family use heat pumps and electric fans for cooling in summer, and mainly kerosene oil heaters for heating in winter. In summer, electric fans are used for longer periods as the fans consume much less electricity at 35 watts, compared with heat pumps (450–600 watts). Two of the three heat pumps were bought 8 years ago. One is in the living-dining area; one in the main bedroom, where the family stays for long periods; and the third one was originally in the child's room when they moved in. The heat pumps are used almost every day between July and September, usually to cool the bedroom for 2 hours before going to bed. They have two oil heaters, one in the living-dining area and the other in the study. These are used almost every day in winter, from November to March, for 5 hours a day: 3 hours in the early morning and 2 hours in the evening. The heat pump is used to warm up the room quickly before the oil heater takes over.

All the members sleep in the larger upstairs room, with mattresses for all on the *tatami* flooring (traditional Japanese mats made of soft straw), which is the typical Japanese way of living. In the daytime on weekdays, they mainly use the downstairs spacious open-plan room, including kitchen, dining and living areas, and the Japanese-style room, with *fusuma* (rectangular sliding walls) between them kept open. Traditional habits and the spatial organisation reduce the number of rooms used by family members at one time, thereby reducing heating and cooling costs.

Around 9 litres per week of kerosene are used for heating. Despite the use of oil heaters in winter, the electricity consumption has been high in winter as well as summer over the past year. This is probably from the use of the heat pump and the mattress dryer, which is on for 1 hour before they go to bed. Kerosene has an energy content of 36.6 megajoules per litre (Wright, undated), so the weekly kerosene consumption is 329 megajoules or about 91 kilowatt hours, 13 kilowatt hours per day during the winter.

The breakdown of the generation of electricity provided by the local electrical company from 1 April to 31 March 2015 is 60 per cent liquefied natural gas (LNG), 24 per cent coal, 6 per cent hydroelectricity, 1 per cent oil, 4 per cent renewable energies and 5 per cent others. Before the 2011 accident at the Fukushima nuclear power plant, caused by the large-scale earthquake, nuclear power accounted for 15 per cent; however, the nuclear power plants have since been stopped for inspection. The reason Fumio and Hana use oil heaters in winter is that they believe it is better for the environment than electricity. One of the major motivations for reducing electricity consumption is their opposition to nuclear power plants.

Natural gas is used for cooking and heating water. In the past year, they used 40 cubic metres of natural gas in December, and only 10 cubic metres in July. This is because in winter they use hot water for washing dishes and in the bathtub, whereas they only take showers with tepid water in summer. The family avoid running water for long periods and using detergent for washing dishes. They also carefully separate their trash for recycling.

Afiq and Mira: A day in Penang: Malaysia

Researcher and writer: Jestin Norden

Family background

This Malay family of six consists of the parents and their three daughters, aged 11, 8 and 3, and a nearly 2-year-old baby boy. Both the husband (Afiq) and wife (Mira) are in their late thirties. They live in the suburb of Gelugor, District of Timur Laut, in the island state of Penang. Based on the latest 2014 population and housing census, there are approximately 529,000 people in Timur Laut. Penang, with a population of approximately 1.6 million, is located in the northern part of Peninsular Malaysia (National Higher Education Research Institute, 2016). It is a modern metropolitan island city that is connected to the mainland via a ferry service (since the 1950s) and two bridges.

Afiq is a member of the teaching staff at a local university, and Mira is a registered pharmacist working with the Penang Health Department. As both are working full-time, their total household monthly income of roughly $US3,760 (i.e. Malaysian ringgit [RM] 15,000) a month is sufficient to maintain their growing family. With this income, the household falls under the upper income-category family in Malaysia (Sulaiman et al., 2015).

What and how much do the family eat?

Since both parents are working, their menu at breakfast is usually a half-boiled egg (two for each parent) with hot tea, or two sausages (boiled, not grilled) with a cup of hot Milo. Sometimes, the two older children have cereal with milk at the table. The breakfasts will be prepared by Afiq as Mira will be preparing the two smaller children's diaper bags. Since breakfasts are prepared at home, the total cost of these meals for six is approximately RM 10. If this family has breakfast outside the home, the cost will simply be doubled.

For lunch, usually the family members each have a plate of plain steamed rice with some greens, and a piece of either fried or curried chicken or chicken soup, or sometimes fish. The reason for chicken always being the first choice is because of the lower price compared with the prices of fish and other seafood. The parents eat separately at their office/workplace, while the children have their lunch at the after-school day-care centre.

Dinner is usually a big thing for the family as everybody sits together at the dining table, shares news and talks about their day at school and at work. Menus for the dinner are almost the same as for the lunches – a cup of rice with either chicken or fish. However, these meals are usually lighter than the meals at lunch time. Cold drinks are preferred at this meal. However, whenever Mira has some extra time in the afternoon after work, she usually prepares the dinner, and this happens at least three to four times a week. The family eat out quite often, at least two to three times a week, because of the busy schedules of the parents, and this is usually dinner. When eating out, however, the family very seldom choose rice as their preferred meal, but will select a lighter meal such as *roti canai*, *parattha* or just a plate of fried noodles. The *mamak* (Indian Muslim) stalls or restaurants would be the first choice for eating out. Once a month, the parents will treat their children to the fast-food chains such as KFC or Subway. At each visit to the *mamak* restaurants, the family will spend around RM 30–35. The total spend is higher by RM 10 when eating out at a fast food outlet.

Knowing that rice is the main source of carbohydrates, the family limit their rice intake to only a cup every time to lower the risk of getting diabetes and growing too fat. Both the parents are already stage 1 overweight (Diabetes UK, 2016).

Snacks between meals

At school, during their 20-minute recess time at around 10 am, the children have a light meal such as a small plate of noodles, or the famous *nasi lemak* (fragrant rice) with anchovies, *sambal* (spicy relish) and cold refreshments. This is considered snacking as the meals are not as heavy as what is served at lunch after morning school finishes at 2 pm. The children will again be snacking on crackers at around 4 pm for their afternoon tea at the afternoon religious classes. Afiq gives each child RM 5 to spend at their school.

The parents, however, do not do snack before lunch at 1 pm, but will have light food such as a piece of curry puff with a *teh tarik* (hot milk tea) or a *kopi-O kaw* (long black coffee) at around 4 pm at tea break. The cost for the afternoon tea snack is always RM 2–2.50.

Snacking after dinner always involves a slice of apple or two, just enough to feed the craving an hour before bed, with some milk or plain water. An apple imported from New Zealand sells for RM 1 at the local hypermarket.

Transportation

This family possesses a compact car, a large people mover/van and a seldom-used small Malaysian-made motorcycle. The final instalment for the car was paid last year; however, the household still needs to pay around RM 1,000 to the bank for the big people mover, which will continue until August 2017. They paid RM 1,100 cash for the used motorcycle.

The two vehicles are used on a daily basis to transport the children to school and to the day-care centre. Afiq commutes in the small car every day to the office after dropping the toddler and baby at the day-care centre. His office is roughly a 3-kilometre walk from the house, but since it is located on the neighbouring hilly suburb of Minden, and given the sometimes unbearable days of tropical heat, he chooses to drive to work in his air-conditioned car.

Mira drives the automatic van roughly 40 kilometres (return) every day to her workplace in the central city of George Town. Before heading to her office, she stops to drop the two older girls at their school just 1 kilometre from their house. Even though there is public transport available in the form of Rapid Penang buses and taxis that connect the suburb of Glugor and the city-centre area, the heavy traffic in the morning and after work always makes the journey very unpleasant. As Mira is unable to arrive late for work, which starts at 8 am every day, she finds that driving is preferable, and her office provides each officer one parking spot.

Almost every fortnight the family will make a trip to visit Afiq's 65-year-old mother. The nearly 90-kilometre journey from Penang up north takes approximately 1 hour. A visit to Mira's parents' house every fortnight is also a must; the return travel distance is only 50 kilometres.

As well as visiting the parents, a trip to the hypermarket to restock the kitchen is a weekly must, and this is usually done on Saturday or Sunday morning and will include a trip to the favourite *nasi kandar mamak* (local Indian rice delicacies) restaurant for lunch. A round trip of approximately 40 kilometres is always made on a weekly basis. This family spend approximately RM 160 a week on fuel for the car, and RM 300 for the van.

Consumer goods

Just like any other ordinary family, this family has at least one of all the basic appliances such as a TV, DVD player, washing machine, iron, second-hand freezer, electric oven, gas burner, microwave, blenders, electric kettle, two one-horsepower air-conditioners (each equivalent to 2.5 kilowatts of cooling), a computer, two laptops, two tablets and two cell phones. All of these appliances are more than a year old.

Recently, Mira bought a new laptop and cell phone as the old ones were no longer functioning well. The new laptop cost her RM 1,200, while the new cell phone cost RM 800. The latest gadgets are not a main necessity for this family.

Energy

Every month, the family uses around 700 kilowatt hours of electricity, and for several months (January to May), the bills do not exceed RM 300. However, last year the electricity bill was the most high for the month of June 2016, when the amount of power used was nearly 1,000 kilowatt hours. The high usage could be due to the heavy demand of the two air-conditioners during the hot season experienced at this time in Malaysia. Both individual air-conditioners are set to switch on automatically at 8:30 pm and off at 5:30 am (9 hours) every day, except on the days when the family are away.

Housing

The family are considered lucky to be able to rent a 2,200-square foot (204-square metre) double-storey terraced house in an established and prime residential area of Gelugor suburb. The unfurnished house is rented for RM 1,250 a month.

The house has four good-sized bedrooms, with three attached bathrooms. Three bedrooms are at the first-floor level, while another is at ground level with its own attached bathroom. There is a smaller room in the kitchen area originally designed for a maid, but this is now used as a store. All bedrooms in the house have timber parquet flooring, except for the lower level and maid's room, which have terrazzo. The split level between the dining room and the living area has created a double-volume living space which makes the interior of the house look bigger. The good-sized 2.4 metre by 2.4 metre internal courtyard between the living room and the ground-level bedroom brings fresh air into the house when the courtyard glass sliding doors are completely open.

The kitchen is attached to a laundry room and a 28-square metre drying yard. Even though the house was built around 30 years ago, the quality of the build is exceptional. The house is made of red clay bricks with cement plaster on the walls, with a reinforced concrete superstructure. The roof of the house is of terracotta clay tiles.

At the moment, the market value of the house is RM 1.28 million (iProperty. com, 2016). This means the house is out of their reach to purchase, even though its location is so appealing with public amenities such as the school, university, wet market, hypermarket, groceries, masjid and healthcare provider all within a radius of just 5 kilometres.

Mongo and Dash: From nomads to town dwellers: Mongolia

Researcher and writer: Fabricio Chicca

The house

The Mongolian family is a couple and one child; the husband Mongo is 33 years old, the wife Dash is 30 and their son Arslan is 7. They live in Ulaanbaatur, to the north of the central district of Zuun ard Ayushin. Their plot is 34 metres long and 15 metres wide, and the house is 8 metres by 7 metres, located at the northern limit of the site. The house has walls of fired brick and the roof is of opaque corrugated

PVC, covered by thick fabric and plastic canvas. Inside, the house is divided into five areas. The main space is where the meals are cooked, and the social part of life happens. In the side opposite the main door which leads from the street, there are three doors leading to a bathroom, the boy's room and a spare room. These divisions were added later, and the family, in case they have another baby, are considering reducing the shared space and creating a new room. Currently, Arslan's room will only fit a small bed, small cabinet and a heater. The parents' room is more spacious and comfortably fits a bed, a large wardrobe and another heater. The bathroom is also reasonably spacious, with a squat toilet, a space for a tap and a bucket underneath. The shared space has a gas stove, coal-burning heater, table, shelves, television, DVD player and stereo, two-seat sofa, sink, food preparation counter, fridge, pantry and desk with a desktop computer and landline phone. The living room is covered with animal skins, plastic and fabric canvas, and has some decoration. There are only four windows in the house, one in the parents' room, two small ones in the living room and the other in the bathroom, although the windows in the living room and bedroom are kept covered by animal skins and canvas.

Services and energy

The house is served by electricity and water supplies. In the past years, the services have been reliable; however, the toilet and kitchen sink are not connected to any reticulated sewerage and the house sewerage and waste water go to a tank and then infiltrate into the soil. During the coldest months, the family spend around tughrik (₮) 18,000 a month on electricity, which is around 272 kilowatt hours per month (1 ₮ = 0.00041 $US). However, during the summer, the bills reduce by about half. In the week before the day of the description, the family had consumed 45 kilograms of coal for heating as the average temperature was −30°C. Although coal is considerably cheaper than electricity, the family has been using the electric oil-filled heaters more often lately, as they felt the benefits of not having coal and ashes inside the house. At this time of year, before going to bed, Mongo loads the coal burner and turns both heaters on for the rest of the night. Before they had the electric heaters, both Mongo and Dash had to share the burden of getting up to feed the coal burner every couple of hours in the night during the winter. Since the temperature in winter can go below −30°C, the heaters are an essential part of daily life.

Daily routine

Dash works part time in a bank while she finishes her university degrees in management and tourism, while Mongo works as a driver. He keeps the car (a medium-sized saloon) from the company with him during the weekdays. The car is used to drive Arslan to school and to drop Dash on her way to university in the morning. On weekdays in the winter, the family routine starts at 5 am, when both Mongo and Dash get up. He goes out and plugs the car heater in so that after a couple of hours he will be able to turn over the car's engine as it will be warm enough. He is also responsible for starting the coal burner and putting the water for bathing on to heat. After that, the family takes their breakfast around 6 am, bathes, dresses and leaves for work and school before 7 am. Mongo leaves Arslan at school, located 1.1 kilometres from home. After that, he drops Dash off and goes to work, 4 kilometres away. Dash then takes a bus to the Mongolian International University 6 kilometres away. The university is a private institution, which the family would not normally be able to afford. However, in 2013, she was awarded a scholarship from the bank where she has been working for the past 10 years. When her classes finish, around noon, she takes another bus to work, 4.4 kilometres away. After work, she walks 1.9 kilometres along Peace Avenue to meet Mongo. During her walk, she stops in different places for groceries. Occasionally, when she cannot find what she was after, she meets Mongo and they go together by car to get groceries near their house. Despite the lack of variety, the prices from the retail outlets near their house are more affordable.

After this occasional quick shop, they pick up Arslan around 5 pm and go home. Mongo drives around 8 kilometres a day going to and coming from work. As his work is as a driver, the amount of additional driving he does during the day varies every week. During the past week, he drove around 300 kilometres. On Friday nights, as he does not have the car, they come back by bus, stopping to pick up Arslan from school. On Monday mornings the entire family goes to work and school either by bus or van (the transport system in Ulaanbaatur is heavily dependent on these illegal but "accepted by the authorities" 15-seater vans).

Both Mongo and Dash share similar heritage and origins. They are both from nomadic families who in the past 20 years have been made to come and live in the town. Mongo came to Ulaanbaatur in 2001 and Dash in 1990. Before they built their current house, they used to live in a traditional demountable hut, locally called a *ger*, which still stands in the middle of the family's site. It is currently used as an additional living space for family and guests during events, and to accommodate their parents when they come to town for a visit. According to the family, they do not live there anymore as the current house has better insulation, requires less maintenance and gives more privacy to the couple. Mongo's parents still keep a nomadic lifestyle, while Dash's parents are now settled in a small village in a mining area by the Chinese border. Like many of the residents in the new neighbourhoods in Ulaanbaatur, the nomadic heritage influences the day-to-day lives of the families in terms of dietary habits, social life and private pursuits. Every night, for instance, while Dash is preparing the food, Mongo goes out for a quick chat with their neighbours. They sit around a fire and tell histories about the times they used to live as nomadic people. This is not exactly a nomadic habit, as the nomadic community was scattered, but each extended family would get together every night to report about the livestock, condition of the grass and the weather. The host neighbour offers tea and some food. According to Dash, Mongolians have a long tradition of hospitality, because if someone comes to the house, it means they have something important to say or do. In nomadic times, a visit meant many kilometres of walking or horse riding. Because of that, tradition requires that visitors have to be well received.

Mongo comes back home for dinner to find Dash either helping Arslan with his homework or finishing the dinner. For the meal, they sit at the table and turn off the television. They would like to be sitting close to the ground, as in traditional dining, but the main door has poor insulation and they can feel a cold draught coming from under it. After dinner, in the winter, they watch television for a couple of hours and then go to bed.

They still follow the habit of eating a single large meal at night, heavily based on goat and cow meat and dairy products, as growing plants for food was not an essential part of the nomadic culture, with small meals in the morning (normally a piece of bread or an egg). Vegetables are not very affordable, and according to Dash, dishes made exclusively with vegetables are often considered second-class food. Despite being strongly based on meat and with few spices available, the family did not repeat a single main meal during the week of this investigation. Mongolians are incredibly creative at mixing milk, meat and some vegetables. During the week, the family consumed 4 kilograms of mutton, 3 kilograms of red meat, 7 litres of milk, 1.5 kilograms of barley flour, 2 large packs of noodles (1.5 kilograms), 400 grams of fresh Mongolian dumpling dough, 500 grams of butter, 250 grams of sugar, 16 eggs, 3 kilograms of animal fat, a block of *aaruul* (cured dehydrated and dried milk), one loaf of bread and six mugs of rice. They also mostly ate traditional dishes, such as deep-fried dough stuffed with meat and onion (*buuz*); baked mutton with rice and onions (*horhog*); thick soup with noodles, meat and fat; a milk soup with meat and eggs; mutton porridge; noodles with *aaruul* and meat; and Mongolian dumplings. During the winter, the majority of the dishes are prepared with animal fat to help the local people handle the harsh conditions. During the investigation, the temperature plummeted to −41°C for three days.

Another perceived influence from the nomadic heritage can be seen in the way the family spends their free time. During summer nights, along with some of their neighbours, they go to the edge of the town to watch the training for the Naadam festival of Mongolian wrestling, horse racing and archery. During the winter, at the weekends, the family go for regular short treks in the mountains close to their house. They also regularly receive friends and family, when the *ger* becomes really useful. They often get together with Mongo's family for regular meals. They meet at least once a week, normally at the weekends. Most of Mongo's family lives in the same neighbourhood, only a short walk away, while Dash's family mostly lives in the old Soviet buildings close to downtown Ulaanbaatur. During these meetings, a meal is normally served, followed by tea, but this did not happen in the week described due to the unusually cold weather. In the school holidays, Mongo sends Arslan to spend time with his parents and learn about the nomadic culture.

Part of the weekend is also used for domestic tasks, such as cleaning the house, repairing the *ger* and washing clothes. Everyone has a specific task; for example, while Dash washes the clothes using the shower bucket and puts them to dry on a line inside their bedroom, Mongo and Arslan are responsible for the cleaning. All of them work together on the *ger* repair and maintenance. They said that Saturday morning is the domestic working day, but they also confessed that they occasionally do everything on Friday night to have more free time at the weekends.

Nanda and Dagian: Life by Lake Inle: Myanmar
Researcher and writer: Fabricio Chicca

The house

The family live in the province of Shan by Lake Inle, in a small house south of Nyaungshwe City. The family consists of a husband (Nanda), wife (Dagian) and two daughters aged 10 and 8 years (Hla and Mya). According to Dagian, their family organisation is common for families that are climbing the social ladder. The family live in a timber house close to the Maing Thauk Monastery. The house is 9.5 metres by 4.5 metres with timber plank walls and a timber structure with a galvanised steel roof. Internally there is a common space where the food is prepared and where the family have their meals, with a small table, four chairs, a wood burner and a gas stove. The rest of the house comprises a small room for the girls and a room for the parents accessed through the girls' room. Every room has two windows, making it reasonably cold. The floor is basically cement and sand. The house is supplied with water and with electricity from the village generator. The supply of water used to be reliable, but lately it is less so. According to the family, as the water from the lake is getting more polluted, the water supply is getting more infrequent. During the period described, the family received water for a couple of hours every other day. For this reason, the majority of the houses in the region have a small water reservoir elevated 3 metres from the ground. Beside the house, 10 metres away from it at the limit of the site, there is a latrine, which is located beside the neighbour's latrine. Both latrines are very close to a creek that flows into the lake. On the opposite side of the house, there is a small structure where the family take showers, and a fenced space for 12 chickens.

Energy

In 2013, the region was connected to the electricity grid; however, the family have chosen to remain connected to the village generator, which has been the reason for some family disputes, as Dagian would like to be connected to the grid, while Nanda does not want this. The generator provides 3 hours of electricity from twilight onwards. The family has three electric lamps, one for each room, and an outlet where a 19-inch television is plugged in and kept permanently turned on. The light bulbs also are always on as they have no switch. Considering the family's electrical demand, the consumption was estimated to be 10 kilowatts a month. The family pay a fixed energy fee, equal to U$10 per month in October 2015. Nanda believes that being connected to the grid will increase the family's expenditure, while Dagian says that a refrigerator would allow her to prepare the meals for more days and store food for longer, as most goes off quickly due to the hot weather.

Daily routine

The family has a very tight routine. Dagian works in Nyaungshwe in a hotel at the reception as she speaks English. The hotel is 13 kilometres from home and she works 6 days a week from 5 am to 3 pm and goes every morning by bicycle. During this week, she went to town six times, cycling approximately 156 kilometres in total. Nanda has a variety of jobs. He has a boat that can be used either for fishing or carrying tourists. His daily routine depends on getting tourists in Nyaungshwe; it starts at 7 am when he warms the breakfast for himself and the girls, and makes sure they are ready to go to school. All of them leave the house around 7:30 am. The girls walk to school, which is 1 kilometre away in the next village, and they each walked 10 kilometres in the week. Nanda takes his boat and goes to the small little wharf in Nyaungshwe where the tourists gather at around 9 am for boat rides. If he manages to get some tourists, the rides take approximately 5 hours, with a trip around the lake, stopping at the monasteries, floating gardens and floating souvenir shopping venues. On trip days he has to come back to the wharf to drop the tourists, and occasionally he picks up Dagian after work as she always checks to see if he is at the wharf before she bikes home. If Nanda fails to get tourists, he has two other possible activities. The first is to take care of the family's floating garden on the other side of the lake, 6 kilometres distant. The second activity is fishing, but this is becoming less common lately. The family plot is divided into two areas, approximately 20 metres wide by 100 metres long, and was inherited by Nanda. Near the plot, the family has a little shed of 16 square metres where Nanda stores his tools and occasionally sleeps when the weather is bad. During the harvesting season, the entire family work on the plot. For this, Nanda comes back home and waits for the girls and Dagian before taking the whole family over by boat. At the time of the investigation, 50 per cent of the garden area was used for rice production, and the remaining equally divided between potatoes and tomatoes. The latter crops are sold to restaurants and hotels in Nyaungshwe. The rice is normally sold to middlemen who sell it on in Mandalay, a city 250 kilometres away. Unfortunately, the family do not have a contract with any middlemen, and as a result, the sales season is very stressful for them. During harvesting, which happens for around a month, every 4 months, Nanda only goes to town to try to get tourists sporadically (around once a week). While transporting tourists is more profitable, it is unreliable, while the crops deliver a steadier and more secure income. They are hopeful that the flow of tourists will keep growing and become more reliable so they can drop the floating gardens.

During the week described, Nanda only managed to have one tourist trip on the lake, on the day of a very popular festival for locals and tourists called Paung Daw Oo. This is a religious festival where a statue of Buddha is delivered to Paung

Daw Oo temple. The remaining weekdays he went to town, but did not have any luck picking up tourists. As a result, he worked on the floating gardens most of the week. Over the week, Nanda used 30 litres of gasoline, which is consistent with the family average.

As the Buddha delivery is an important religious event, the family decided to join it. This required a great deal of planning. On the morning of the event, around 5 am, Nanda brought Hla, Mya and Dagian, who was on her day off, to the temple located 14 kilometres to the south, and then rushed to town to get tourists. During the special events, all the boats in the region go to the pier to pick up tourists, and when there are not enough boats, the boatmen are able to increase the ride price. A normal boat ride is around U$10 per person, but during special events, the price may go up to U$50 per person. Nanda eventually picked up four German tourists who paid U$35 each for the ride, representing 30 per cent of the monthly family average income. By the end of the event, at around 1 pm, Nanda brought the tourists back to town, went back to the temple to pick up his family and then travelled back home. The entire journey consumed 12 litres of gasoline, in his 13 horsepower boat engine. During part of the day, the wind was very strong, causing a substantial increase in fuel consumption.

The main entertainments of the family revolve around the lake, religious events or village gatherings. At least once a week the entire family tries to visit the local temple, which is 300 metres from their house. In front of the temple, there is a patio where families gather for a local chat almost every night. Dagian and Nanda go to these informal gatherings almost every day after work. During the week described, they went there for 5 out of 7 days, for around 20 minutes. Around the lake are several large monasteries and each one has its own festivities. According to Dagian, there are more than 20 religious events in the region, but realistically they only manage to go to five or six events every year. It is normal to have several stalls selling food and other goods around the larger temples during the festivities.

During the festival, after the tourists were brought to the temple, the entire family had an opportunity to enjoy the party. The parents had chicken stew and rice, while the daughters had popcorn, guava and soft drinks. Nanda bought himself a new torch and batteries and Hla and Mya had new shoes, dresses and socks.

Food

Eating is an essential part of the family's organisation and lifestyle. Every night Dagian prepares food for the next day's breakfast and a snack for the children and Nanda for the next day's lunch. The family eat together every night while the lights are on.

During the week described, the family were having a simplified version of the traditional *mohinga*, which is a noodle soup with spices and eggs. They had it every day in the morning. Except for the noodles that are bought in town, the family produce everything else. A small space in between the main production areas of the floating gardens is used for herbs, spices, pepper, spinach, ginger and onions and the family is self-sufficient in these products. The girls' snack consists of guavas and mangos, and rice with eggs every day. Nanda has the rice with eggs but not the fruit. Dagian is offered meals at the hotel where she works but she skips these most days and normally only has fruit. During the week described, she only had lunch once, chicken soup with green beans, as it is her favourite dish. All other days she had guavas. At night, the family has variations of the traditional *mohinga*. Three out of the seven nights, the *mohinga* included eggs and small pieces of chicken bought by Dagian on her way back from work. In exchange for a lift from town back to the village, Nanda was given two fish that were eaten that night by the family. One night they had noodle soup with the leftovers from the fish (head and tail cooked with vegetables, coriander and coconut milk). Two nights the family had rice with vegetable stew. The family have a wide selection of fruit available, either from their garden where, depending on the season, they find papaya, guava, mangos and limes, or from their neighbours.

Note

1 Information about the community came from Slamet.

References

ADB and WWF (2012) *Ecological footprint and investment in natural capital in Asia and the Pacific*, available at http://www.footprintnetwork.org/content/images/article_uploads/ecological-footprint-asia-pacific-2012-final.pdf, accessed 29 March 2017.

Aweto A (2013) *Shifting Cultivation and Secondary Succession in the Tropics*, Wallingford, UK: CAB International.

Bolchover J, Lin J and Lange C (2016) *Designing the Rural: A Global Countryside in Flux*, e-book available at http://ebookcentral.proquest.com/lib/VUW/reader.action?docID=4819964, accessed 11 April 2017.

Caldieron, J. and Miller, R. (2013) Residential satisfaction in the informal neighborhoods of Ulaanbaatar, Mongolia, *The ARCC Journal of Architectural Research*, 7(1), pp. 12–18.

Census Population (2015) *Khajuraho Population Census 2011*, available at http://www.census2011.co.in/data/town/802141-khajuraho.html, accessed 31 March 2017.

Central Intelligence Agency (CIA) (n.d. a) *The World Factbook: Urbanisation*, available at https://www.cia.gov/library/publications/the-world-factbook/fields/2212.html, accessed 31 March 2017.

Central Intelligence Agency (CIA) (n.d. b) *The World Factbook: Population growth rate*, available at https://www.cia.gov/library/publications/the-world-factbook/rankorder/2002rank.html, accessed 31 March 2017.

Climate-data (n.d.) *Climate: Nagoya*, available at https://en.climate-data.org/location/4900/, accessed 20 June 2017.

Diabetes UK (2016) *New research claims link between white rice and type 2 diabetes*, available at https://www.diabetes.org.uk/About_us/News_Landing_Page/New-research-claims-link-between-white-rice-and-diabetes/, accessed 3 July 2016.

Esri (2016) Average household size in India, available at http://www.arcgis.com/home/item.html?id=6cf22970ea8c4b338a196879397a76e4, accessed 20 Feb 2017.

Fern K (2014) *Useful tropical plants: Sapindus rarak*, available at http://tropical.theferns.info/viewtropical.php?id=Sapindus+rarak, accessed 8 March 2017.

Global Footprint Network (GFN) (2016) *National Footprint Accounts, 2016 Edition*, available at http://www.footprintnetwork.org/licenses/public-data-package-free-edition/download/, accessed 30 March 2017.

Hamidi J (2016) Modifying the development model for an inclusive museum to realise a miniature of good village governance (a study on indigenous people of Tengger and Baduy), *Pertanika Social Sciences and Humanities*, 24(2), pp. 721–736.

Hara T (2015) *A Shrinking Society*, SpringerBriefs in Population Studies, e-book, DOI 10.1007/978-4-431-54810-2_2.

Htwe T N, Kywe M, Buerkert A and Brinkmann K (2015) Transformation processes in farming systems and surrounding areas of Inle lake, Myanmar, during the last 40 years, *Journal of Land Use Sciences*, 10(2), pp. 205–223.

iProperty Group Limited (2016) Terrace/Link/Townhouse for Sale in Minden Heights, Penang, *Malaysia's No. 1 Property Website*, accessed 23 June 2016.

Jain D K (1990) Impact of tourism on Khajuraho, India: A preliminary analysis, *Tourism Recreation Research*, 15(1), pp. 43–44.

Joshua Project (2017) *Baduy Kanekes in Indonesia*, available at https://joshuaproject.net/people_groups/10549/ID, accessed 7 March 2017.

Kamata T, Reichert J, Tsevegemid T, Kim Y and Sedgewick B (2010) *Best Practices in Scenario-Based Urban Planning: Managing Urban Expansion in Mongolia*, Washington, DC: The World Bank (e-book), available at http://elibrary.worldbank.org/doi/pdf/10.1596/978-0-8213-8314-8, accessed 10 April 2017.

Khajuraho Tourism (n.d.) *Weather and climate of Khajuraho*, available at http://www.khajuraho-india.org/khajuraho-weather.html, accessed 4 April, 2017.

Krishnaswamy S, Subramaniam K, Indran T and Low W-Y (2012) The 2004 tsunami in Penang, Malaysia: Early mental health Intervention, *Asia-Pacific Journal of Public Health*, 24(4), pp. 710–718.

Masron T, Yaakob U, Ayob N M and Mokhtar A S (2012) Population and spatial distribution of urbanisation in Peninsular Malaysia 1957 – 2000, *Malaysia Journal of Society and Space*, 8(2) pp. 20–29.

Merriam-Webster (2007) Asia, *Merriam-Webster's Geographical Dictionary*, available at http://search.credoreference.com/content/entry/mwgeog/asia/0?searchId=d210c1b6-1334-11e7-b19f-0e58d2201a4d&result=0, accessed 28 March 2017.

Myanmar Travel Group (2016), *Climate*, available at http://www.myanmartravelgroup.com/about-myanmar/climate/, accessed 24 April 2017.

National Higher Education Research Institute (2010) The State of Penang, Malaysia: Self-Evaluation Report, *OECD Reviews of Higher Education in Regional and City Development*, IMHE, available at http://www.oecd.org/edu/imhe/regionaldevelopment, accessed 5 April 2017.

National Tropical Botanical Garden (2017) *Etlingera elatior*, available at National Tropical Botanical Garden, 2017, accessed 8 March 2017.

Ota A (2006) *Changes of Regimes and Social Dynamics in West Java*, Leiden; Boston, MA: Brill.

Penang Institute (2016) *Population*, available at http://penanginstitute.org/v3/resources/data-centre/122-population, accessed 3 July 2016.

Sidle R C, Ziegler A D and Vogler J B (2007) Contemporary changes in open water surface area of Lake Inle, Myanmar, *Sustainability Science*, 2(1), pp. 55–65.

Statistics Bureau (Japan) (2016) *Statistical Handbook of Japan 2016*, available at http://www.stat.go.jp/english/data/handbook/c0117.htm#c02, accessed 20 June 2017.

Su M and Jassby A D (2000) Inle: A large Myanmar lake in transition, *Lakes and Reservoirs: Research and Management*, 5(1), pp. 49–54.

Sulaiman N, Fauzi S F and Saukani M N M (2015) Impact of Social Capital on the Household Income Group in Malaysia. Prosiding Persidangan Kebangsaan Ekonomi Malaysia Ke-10 2015, 2015. Kuala Lumpur, pp.74–84.

United Nations (2015) *World Urbanization Prospects: The 2014 Revision*, New York: United Nations Department of Economic and Social Affairs.

United Nations Educational, Scientific and Cultural Organization (UNESCO) (2017) *Khajuraho Group of Monuments*, available at http://whc.unesco.org/en/list/240, accessed 31 March 2017.

University of Iowa (n.d.) *The Baduy of Indonesia*, available at https://clas.uiowa.edu/linguistics/baduy-indonesia, accessed 7 March 2017.

Vale R and Vale B (2009) *Time To Eat the Dog? The Real Guide to Sustainable Living*, London: Thames and Hudson.

Vale R and Vale B (2013) Domestic travel in Vale R and Vale B (eds.) *Living Within a Fair Share Ecological Footprint*, London: Earthscan.

Wan G and Wang C (2014) Unprecedented urbanisation in Asia and its impacts on the environment, *The Australian Economic Review*, 47(3), pp. 378–385.

World Bank Group (2016) *Urban population (% of total)*, available at http://data.worldbank.org/indicator/SP.URB.TOTL.IN.ZS, accessed 31 March 2017.

World Weather and Climate Information (2016a) Average monthly weather in Penang, Malaysia, available at https://weather-and-climate.com/average-monthly-Rainfall-Temperature-Sunshine,Penang,Malaysia, accessed 6 April 2017.

World Weather and Climate Information (2016b) Average monthly weather in Ulaanbaatar, Mongolia, available at https://weather-and-climate.com/average-monthly-Rainfall-Temperature-Sunshine,ulaanbaatar,Mongolia, accessed 10 April 2017.

World Weather Online (n.d.) *Khajuraho, Madhya Pradesh monthly climate average, India*, available at https://www.worldweatheronline.com/khajuraho-weather-averages/madhya-pradesh/in.aspx, accessed 4 April 2017.

Worldometers (2017) *Asia Population*, available at http://www.worldometers.info/world-population/asia-population/, accessed 28 March 2017.

Wright M (undated) *Fuel energy conversion factors*, Berkeley, CA, University of California, Astronomy Department, available at http://w.astro.berkeley.edu/~wright/fuel_energy.html, accessed 26 August 2017.

4
EUROPE

Adele Leah, Hannu I. Heikkinen, Han Thuc Tran, Ludwig Geisenberger, Stefan Opfermann, Robert Vale and Brenda Vale

The continent of Europe

Brenda Vale

In 2017, Europe was home to 740 million people. It was also home to the 800 people who live in the Vatican City, the smallest country in the world (Worldometers, 2017), with an area of 44 hectares. Europe includes the other very small countries of San Marino, Lichtenstein and Monaco, all with populations below 50,000. The largest country in Europe is the area of Russia west of the Ural Mountains, followed by Ukraine. Apart from the very small European countries, which are densely populated because they are very small, the Netherlands and Belgium have the most people in the least space, as in 2013 there were 498 and 370 people per square kilometre, respectively, living there (StatisticsTimes, 2016). The three stories in this chapter come from Finland (18 people per square kilometre), Germany (231 people per square kilometre) and England, which if taken alone, apart from the rest of the United Kingdom, is more densely populated than the Netherlands. Although this is a generality, national average ecological footprints tend to fall moving from western to eastern Europe.

This section sets the scene with some background information on the places the stories come from.

Oulu, Finland

The largest city in northern Finland, with a population of 200,000, Oulu was once known for its wood tar and salmon, and is now known for its IT industries (BusinessOulu, n.d.). Oulu is sited towards the top of the Gulf of Bothnia off the Baltic Sea and developed around the mouth of the Oulujoki River. The city lies south of the Arctic Circle and has a climate where summers are mild and winters cold with guaranteed snow.

The city has a modern appearance, having grown in recent years with new and refurbished residential areas (City of Oulu, n.d.). This planning means it has a central pedestrian area and an extensive network of cycleways. Despite winter snow, cycling remains an important method of getting around the city, with a modal share of 32 per cent in summer and 12 per cent in winter (Swanson, 2016). At least cycling in winter keeps you warm and, as Swanson notes, it tends to be safer as cars have to go slower in winter to avoid skidding. In Oulu, there are also extensive parks, seaside walks and hiking trails in the surrounding countryside (City of Helsinki, n.d.). The relationship with the countryside is an important part of Finnish culture.

Eschenlohe, Germany

Eschenlohe is a small German municipality in southern Germany with the village of the same name at its centre. It lies close to the border with Austria. Eschenlohe lies on the Loisach, a river that flows from the Tyrol in Austria into Germany. The village has fields around it but wooded hillsides slope up on either side of the valley floor. A new covered bridge crosses the Loisach in the centre of the village (Kaden, 2008). Many houses have the traditional shallow sloping roof and balconies overlooking the street that typify housing in this area.

The climate is mild. July is the hottest month, with a monthly average temperature of 17.5°C, and January has the lowest monthly average of −1.6°C. Maximum temperatures in June, July and August are in the low 20s. It rains throughout the year, but the summer months are the wettest (Climate-Data, n.d.). In August 2005, there was a bad flood, part of severe flooding across Europe, when the Loisach burst its banks after heavy rainfall (European Space Agency, 2017).

Newton-le-Willows, England

Newton-le-Willows is a market town in northwest England with a 2015 estimated population of just over 23,000 (Citypopulation, 2016). Passing through the town is not only what has been claimed as the earliest canal (Rolt, 1950), but also the earliest railway, and with it the first railway accident (Leigh et al., 1830). The railway led to the town's industrialisation as it became known for making railway wagons and locomotives (Anon, 2014). This industrial heritage has now become a tourist attraction (St. Helens Council, 2017).

The climate is warm and temperate with rain throughout the year. The warmest month is July, with a monthly average temperature of 16.3°C, and the coldest January, with a monthly average of 4.2°C (Yr, 2017).

Onni and Leena: Fishing, hunting and berry picking: Finland

Researchers and writers: Hannu I. Heikkinen and Han Thuc Tran

Background

Post-World War II, and especially during the 1960s, Finland experienced rapid industrialisation, urbanisation and changes in consumption habits that saw these become dominated by industrial goods. As a result, in the 2000s, the typical Finnish family live in a city, town or nearby suburb, even though many still belong to the first or second generation of urbanites. Typically these new urban dwellers have relatives or an original homestead somewhere in the Finnish countryside, which often today serves as a holiday destination for families. At the summer cottage, they behave more traditionally, making use of the sauna and taking part in traditional hobbies such as fishing, hunting and berry picking (Vepsäläinen et al., 2015; Official Statistics of Finland, 2014; Heinonen and Peltonen, 2013: 10–37, 254–259).

Our sample family belong to the higher educated middle class and live in a detached house in the city of Oulu, Northern Finland. The yearly average daily temperature is 2.7°C, the lowest average is −13.6°C and the highest average is 20.9°C (Pirinen et al., 2012). The family's home lies some 20 kilometres from the city centre in a historic agricultural municipality, which together with most of the surrounding communities, joined the city of Oulu in 2013, making the region another example of urban sprawl.

Food and eating habits

The family consist of Leena, Onni and their 20-year-old daughter Milla and 19-year-old son Veeti. Both Milla and Veeti have recently moved out to pursue their education and live as singles, along with 48 per cent of Finns (Official Statistics of Finland, 2015). Education takes time in Finland and 40 per cent of those between the ages of 20 and 29 are enrolled in education (OKM, 2016). However, the adult children visit home regularly during holidays, and this serves as the timeframe for this everyday narrative. Both of them have recently turned to veganism or vegetarianism, which has become a trend in Finland among the middle class. The largest group of vegetarians seems to be young females living in cities (Vinnari et al., 2010).

Onni and Leena, 46 and 47 years old, prefer a mixed diet, especially fancy dairy products, but vegan foods have tended to dominate joint meals with the children. Due to the increasing popularity of a vegan diet, imported food products, such as soy, beans, olive oil and palm products, have become more common as well as different kinds of nuts. Domestic sources of plant-based protein have been increasing, but their availability is still rather poor and vegetarian products are expensive. In general, the cost of a healthier diet has been recognised as a problem in Finland (WHO, 2015; Reher, 2013). In any case, the bulk of vegetables and fruits must be imported due to Finland's subarctic farming conditions (Trading Economics, 2016). To acknowledge their agrarian cultural heritage, but also to compensate for the increasing amount of imported food, the family have a small garden for home production of vegetables during the summer months. The small plot grows herbs, salad stuff, onions, strawberries, apples and different currants. Their own and the grandparents' gardens produce vegetables and salads sufficient for 1–2 months of domestic consumption. Almost every month the grandparents provide wild fish for home consumption. Garden-grown berries meet half the yearly consumption and the other half consists of wild berries, such as bilberries and lingonberries, picked from July to September. Most of the mushrooms are also wild, picked from July to September. In general, wild berries and mushrooms have been and still are a very typical speciality in the Finnish diet (Arctic Flavours, 2016).

In our sample week, the common breakfast begins with lingonberry juice or orange juice and coffee. Typically, all family members have their own version of breakfast. Onni and Leena eat one to three pieces of rye, barley or oat bread with cheese, garlic and salad on top. Leena prefers poultry cold cuts with this and Onni liver sausage. Milla and Veeti prefer muesli or cereals with frozen bilberries or blackcurrants and vegan milk substitute. Usually, imported fruits are eaten as well, such as bananas, apples or grapes. During the daytime, Leena prefers to take a small packed lunch with her, while the others have small snacks consisting of different fruits, tomatoes and root vegetables. Typically, Onni has carrot, apple, banana and tomato with him at work. All have their main dinner together after work or school at around 6 pm. In our example week, on Saturday, salmon and marrow bake was the main dish and on Sunday, fried vendace (a local fish) with baked potatoes. The remains of these dishes were consumed until Tuesday. On Wednesday evening, a big pot of root vegetable soup was prepared (beetroot, turnips, carrots, onions, garlic and chilli) and it was eaten until Friday night.

Work and school days are long and it is common in work weeks that dishes are prepared during the weekend and warmed up in the evenings. At the weekends, Onni and Leena have red wine with dinner and beer is consumed when they take a sauna. Leena smokes daily. A typical dinner is homemade vegetarian pasta with a tomato and bean sauce, with at times added soy, beans or oat-based meat protein substitute. Cooked or baked potatoes and other root vegetables are common as well. Onni and Leena eat beef, wild or farmed fish, reindeer or game two or three times a week, and dairy products daily. The most common evening snack consists of salted or natural nuts. Lunch or dinner in a restaurant is rare, only happening once a month or so.

Transportation

Onni works in the city of Oulu and commutes 50 kilometres daily in a 1.4-litre diesel car. During the five warmest months he uses a 535 cc motorbike. Leena works in the nearby library and walks or cycles 2 kilometres on a daily basis. Onni does four to eight longer domestic or foreign work trips a year, either flying or travelling by train (>1000 kilometres). Leena also does work trips but these are local (50 kilometres) and she uses the bus. Both Milla and Veeti use bicycles to get to school, going around 10 kilometres daily, but at times they also use public transport. The family have six bicycles and four are used daily. They like to have some physical exercise every other day to avoid the dangers of sedentary lifestyles (Lista, 2015; Valtonen et al., 2010). Their most common exercises are jogging, rollerblading, skiing or visiting the nearby gym, which all take place within walking distance. During the eight cold months, Onni goes by car to play ice hockey once a week in the nearby hockey arena (16 kilometres). For their music hobby after work, Onni uses the car and Veeti his bicycle for rehearsals (10 kilometres) once a week. Either or both children are taken once a week to their apartment or to the railway station by car (50 kilometres). Buying groceries is done by bicycle or walking (500–1,000 metres). The family have a tradition of making a visit once a month to the grandparents and relatives in Onni's rural birthplace (550 kilometres), but other trips are rare, with a trip abroad maybe every other year.

Consumer goods

The family have a TV, VHR and CD player, radio and stereos, as well as three desktop computers, two laptops and four smartphones. The parents have a dishwasher, washing machine, blender, refrigerator, two freezers, microwave, stove, wood-heated oven, sauna and fireplace. All major gadgets are 5–15 years old. Both adult children live in separate housing during school time, and Veeti has a dishwasher, washing machine, blender, refrigerator, freezer, microwave and stove. Milla used to live in the same apartment during the time she was at high school, but has now moved to another city for university and today has access to the consumer goods at home and the communal equipment of a dormitory. Due to their music hobby, the family have two acoustic guitars, a self-made electric guitar, electric bass, keyboards, drums and a couple of minor instruments. They also have four pairs of roller blades, skis and skates for sports and a collection of hundreds of movies on discs and tapes.

Energy

Energy consumption of the main dwelling is estimated to have been 11–12,000 kilowatt hours a year, but this has dropped to 9,877 kilowatt hours per year since the children moved out for their schooling. The yearly estimated energy consumption of the children's apartment is 917 kilowatt hours (plus district heating). The electricity provider produces electricity from hydro, wind and peat-burning power plants. Oil and coal might be used during extremely cold winter days. The main house has a wood-burning oven, fireplace and sauna stove which together consume some 6–8 cubic metres of stacked firewood every year. The sauna is heated almost every evening and the wood-fired oven is lit every other day during the six coldest months. Firewood is acquired yearly as a side product of the thinning of the family's forests, owned and co-managed by Leena's family.

Housing

The detached house was built in 1969 of red brick with a wood frame and concrete footings. The heated part is 120 square metres, while the cold storage/garage part is 60 square metres. The house consists of an outdoor terrace, four bedrooms, lounge, kitchen, utility room, shower and sauna. Major renovations were made in the 1980s, as well as in several phases during the 2000s. Today the roof is made of steel, windows are multi-glazed and doors have good insulation. The original insulation is glass-wool based, except for the ceiling, which is insulated with planing-mill sawdust. Extra ceiling insulation was added in 2002, in the form of recycled paper. The house uses geothermal heating, which also gives daily hot water. The house is connected to the municipal infrastructure and the estimate for water consumption and corresponding waste water has averaged some 112 cubic metres a year. After the children moved out for schooling, this dropped to 75 cubic metres. The water/wastewater consumption estimate in the children's apartment is 30 cubic metres of cold water and 21 cubic metres of hot water, calculated separately because of the district heating system. The trash can for the family's house holds 240 litres and is emptied every fifth week. Recycling collection bins are adjacent to the nearby grocery store and are used while shopping.

Heinrich and Albrecht: "Finally there are children living in our street again": Germany

Researchers and writers: Ludwig Geisenberger and Stefan Opfermann

Background

Heinrich and Albrecht, with sometimes a temporary dog (at the moment it is a boxer) as one of them is a vet, live in Eschenlohe, a village with about 1,600 inhabitants and currently 20 "refugees" from Syria and Afghanistan who live in a neighbouring house. As Heinrich said, "Finally there are children living in our street again."

Eschenlohe is about 80 kilometres from Munich by motorway. Eschenlohe has a grocery shop; a baker; a butcher; a shop that sells all kinds of drink; a bicycle shop; three inns which offer food, drink and accommodation; a railway station; an infants' school and two public telephone boxes. To the nearest larger town of Murnau for shopping or the doctor, it is around 9 kilometres. Anything that cannot be found there can be obtained in Munich.

House and transport energy

The house, which was inherited from parents, has six rooms and stands on a plot of about 500 square metres. Heinrich and Albrecht heat with oil (about 1,200 litres per year) and with wood (3 *Ster* per year—a *Ster* is 1 cubic metre, so this is 3 cubic metres of firewood). Of course, the house has electricity, water and telephone and internet connections, but there is no television and many books instead. The kitchen has the usual stove, fridge and freezer plus a number of small appliances (electric slow cooker, kettle, toaster, stick blender, cream beater, small grain mill). There is a washing machine, two radios and a computer, printer and shredder. Tools consist of a drill, jigsaw, planer and a large wood splitter, together with a lawn mower, battery charger for the car and chargers for the two cell phones.

Heinrich and Albrecht have a middle-sized efficient car which uses 5 litres of diesel per 100 kilometres and is driven roughly 30,000 kilometres per year. Twice a year they visit a 93-year-old aunt in Budapest and they visit relatives in other parts of Germany.

Routine

- A typical Saturday is listed chronologically below. All timings are given as approximate values.
- Get up between 6 and 7 am.
- 30 minutes' morning meditation.
- 30 minutes' exercise (for the back).
- Grind flour for bread (using an electric grinder) and mix dough.
- Breakfast.
- 45 minutes walking the dog.
- Go shopping in Murnau (population 12,000), about 9 kilometres distance.
- Cook lunch of three separate vegetables (usually roots, leaves and fruit) with millet or potatoes.
- Lunch followed by a quite important mid-day sleep.
- 45 minutes walking the dog.
- Cleaning the house—this is not done very often so it takes a long time.
- Baking bread with baking powder because one person cannot tolerate yeast of sourdough.
- Evening meal—this is usually a light meal with bread. Doing homework for two anthroposophical working groups or else preparing this.
- To end the day, reading a novel or something similarly light.
- 30 minutes' evening meditation.

Electricity

As well as the oil and wood already mentioned, Heinrich and Albrecht use about 3,900 kilowatt hours of electricity a year. Comparatively, this is a lot of electricity for a household of two people, but it is explained by the fact there is a sauna that is normally used twice a week. The electricity is so-called *Öko-Strom* (Eco-power), which is generated from water, sun and wind energy with no use of nuclear energy and no CO_2 emissions. They pay 26.22 Euro cents per kilowatt hour.

The house

The house of 125 square metres was built in 1964. The external walls are made of pumice blocks and the internal walls are brick. The house was insulated about 15 years ago with polyurethane foam on the ground floor and in the roof and mineral fibre on the first floor. The windows on the south side are poorly insulated but those on the other sides of the house have well-insulated glazing. In winter, the south windows are closed off with wooden shutters at night.

On the roof there is a 12-square metre thermal solar collector. In summer, this provides enough heat for hot water and to keep the bathroom floor warm. Any heat collected in winter helps to pre-heat the hot water supply. A low-temperature heating panel in the wall provides supplementary heating, and most of the year heat from this is sufficient. The heat from the wood-burner provides additional heat for the living room.

Travel

The trips to Budapest (about 700 kilometres) are made by train. Trips in Germany are mostly by car and only sometimes by train.

Food

As far as possible, food products come from regional ecological production with the exception of things like sugar and coffee. On weekdays, Heinrich and Albrecht cook vegetarian food mornings and evenings, but one has some sausage. On Sundays, they often eat out and then one definitely eats meat and the other sometimes.

Hester and Julian: The costs of car travel: England

Researcher and writer: Adele Leah

Background

The family live in Newton-le-Willows, a small market town situated in northwest England, equidistant between the cities of Liverpool and Manchester. The daily family routine was recorded in a diary which was completed by Hester.

The house

The family have three members (Julian and Hester, and their young child). The family own their home which they purchased in 2014, shortly after it had been constructed. The house is a two-storey, mid-terraced house, of brick cavity wall construction with a timber-framed roof and clay tiles, which is the usual method of construction for this area of England. The cavity wall is insulated, as is the roof and under the ground floor. The floor area of the house is approximately 75.4 square metres. The ground floor contains a kitchen/dining room, lounge and bathroom. The main entrance to the house is from the street, and the front door opens into a vestibule which gives access to the lounge and the bathroom, which contains a toilet and basin. Linking the lounge and the kitchen/dining room is a hallway which contains the staircase leading up to the first floor.

The kitchen/dining room situated to the rear of the house receives direct sunlight during the morning and afternoon. It contains kitchen units and appliances at one end, and a dining table and chairs at the other. The dining area has double doors which open out into a small, private rear garden. The lounge is situated at the front of the house, looking on to the street, and contains a bay window through which the room receives direct sunlight in the evening. The lounge is well utilised by the family throughout the day, as they move between this and the kitchen/dining room.

The first floor contains three bedrooms, a storage cupboard and a bathroom, all of which are accessed from the first-floor landing. Hester and Julian sleep in the main bedroom, which contains a double bed and built-in wardrobes. A small en-suite containing a shower, wash hand basin and toilet is accessed from this bedroom. Bedroom two contains a double bed, and is reserved for overnight guests. This bedroom is also used for drying washing inside when the weather prevents outside drying. The third bedroom is much smaller and is used as a nursery/bedroom for the child. It is only large enough to accommodate a single bed and no other furniture, and Hester feels this bedroom would be better utilised as a home office/study space. The main bathroom contains a bath with a shower over, a toilet and a basin. The roof space is accessed through a hatch with a drop-down ladder installed in the ceiling of the first-floor landing. It is used to store light items close to the access hatch, as no boards have been laid across the ceiling joists.

The house is serviced with water, sewerage, electricity and gas. There are also connections for telephone, broadband internet and cable television. It is heated by central heating, burning natural gas in a boiler, which serves radiators in all rooms including the vestibule, hallway and ground floor bathroom. The gas central heating is utilised during autumn and winter. A thermostat means the heating system automatically switches on when the temperature of the house drops below 18.5°C. The house heats up very quickly and retains heat well, in part due to it being a mid-terraced house, insulated on each side by the neighbouring houses, but also because of the good levels of insulation. The water is heated by the gas "combi boiler", which only heats the water when a hot tap or shower is turned on. This means that no space is needed to accommodate hot water storage and there is no heat loss from a large tank of hot water. There are thirty-six electric sockets in the house, most of which are located in the kitchen/dining room, lounge and bedroom one. The kitchen contains a fridge, freezer, washer/dryer, dishwasher, electric oven and gas hob, which were all provided with the house. The family were given a microwave and have recently replaced their vacuum cleaner, kettle and toaster. Other electric items in the house include a television, digital radio, stereo, laptop, iPad, two Kindles, four iPods and two mobile phones, all of which are more than 3 years old. The family purchase their gas and electricity from the same energy supplier, and pay a monthly direct debit of GBP 65 throughout the year. There is usually credit in their account at the end of the year.

Household and garden waste are collected by the local council. Recyclable waste of cardboard, paper, glass, tin cans, plastic bottles and food waste is collected from the kerbside each week. The council provides an individual bag/box for each of these types of waste. The food waste is not strictly compostable, as it includes food leftovers, fruit and vegetables, tea bags, dairy products, fish, rice, pasta, beans, meat and bones and stale bread, cakes and pastries. Non-recyclable household and garden wastes are collected once a fortnight. The rear garden was grassed over when the family originally purchased the house, but since then they have replaced the grass with gravel. They did this because of the poor quality of the grass, which became waterlogged after heavy rain, and the earth beneath which they discovered contained building rubble. The family are now recreating the garden with potted plants, seed boxes and climbing plants growing up lattices attached to the fence. Hester and Julian tend the plants while their young child plays in the garden, which contains a small sand pit. There is a washing line for drying clothes outside when the weather permits, and the double doors which connect the dining area with the garden are open most days during spring and summer. The garden receives full sun for most of the day during the summer months. To the front of the house, adjacent to the street, there is a hedge and a small strip of grass, which separates the lounge bay window and the driveway. Hester explains that they would grow fruit and vegetables if their garden was bigger:

> My family home has a big garden with fruit trees (apples and plums) and we had redcurrant bushes and brambles, rhubarb, and hens when we were very little. We grew up picking berries in the summer and plums and apples in the autumn to make jam and chutney. I also remember picking pea pods with our next door neighbour in his garden – and loving the taste of the fresh peas! If we don't have a garden big enough to grow fruit and vegetables in the future I like the idea of having an allotment or being part of a community allotment/gardening scheme.

Food

Food is purchased weekly by Hester from a supermarket, but occasionally milk and bread are purchased from the local shop. The family do not shop at the local supermarket; they consider it to be expensive and also question its business ethics. Consequently, Hester and the child travel by car to another supermarket approximately 8 kilometres away, where they also purchase fuel for their car (which is cheaper than the fuel at their local petrol station). They often combine this trip to the supermarket with a walk in the park.

The family do not have a meal plan for the week, but a typical menu is given in Table 4.1. Dinner is most often the larger meal, and the family eat a lighter lunch. Hester is vegetarian, while Julian and their child are meat eaters.

The family occasionally eat snacks between meals, typically biscuits, fruit, nuts, crackers and bread/toast. Hester prepares most of the food, spending about half an hour to an hour each day cooking. Often, they will eat the same meal two nights together (if there are leftovers), and typically Hester prepares food at the weekends for the week ahead. The family rarely eat takeaway food, but occasionally eat out at an Italian restaurant, and often visit a café for tea/coffee/cake.

TABLE 4.1 Typical weekly menu

	Breakfast	Lunch	Dinner*
Day 1		Sandwiches OR	Pasta with bolognaise/ pesto/another sauce
Day 2		Soup	Risotto
Day 3		OR	Fish and chips
Day 4	Cereal and toast	Crackers and cheese Fruit and/or yoghurt	Casserole and baked potatoes
Day 5			Omelette, chips and salad
Day 6			Baked potatoes with cheese/baked beans/ tuna/salad
Day 7			Quiche, potatoes and salad

*The family occasionally eat dessert with dinner (i.e. some sort of cake or pudding) but usually have fruit and/or yoghurt.

TABLE 4.2 Typical monthly car usage and CO_2 emissions

	Distance covered in a round trip (kilometres)	Frequency of trip (per month)	Total number of kilometres travelled
Travel locally to nursery, toddler groups, to visit friends and for food/other shopping	12	40	480
Travel further afield for recreational purposes	28	6	168
Travel to visit family in York	280	1	280
Travel to visit family in Sheffield	212	0.5	106
Travel to visit family in Ilkley	208	0.5	104
Total number of km travelled in 1 month			1,138 kilometres
Petrol consumption (7.2 litres/100 kilometres) per month			82 litres
CO_2 emissions (173 grams/kilometres) per month			197 kilograms

Transport

Julian works 4 days a week, and Hester 1 day a week, thus sharing childcare. Both commute to work by train on a journey that takes 20 minutes. The railway station is a 20-minute walk away from their house.

The family own one car, with a 1.6-litre petrol engine. This is used multiple times each day for short journeys to destinations typically less than 8 kilometres away from home. These include travel to nursery, toddler groups and activities, to visit friends, and for food/other shopping. Approximately twice a week, the car is used for recreational trips to destinations further afield, but less than 16 kilometres away from the house. Every 2–3 weeks, the family travel in their car to visit family members further away. They do not have a strict travel plan for each month, but typical journeys and distances travelled by car are given in Table 4.2.

The family's car uses approximately 7.2 litres of petrol per 100 kilometres. City driving uses 9.4 litres per 100 kilometres, while motorway driving uses less (6 litres per 100 kilometres). The associated CO_2 emissions are 173 grams per kilometre (Autoevolution, 2017).

Since purchasing their car 2.5 years ago, the family have driven approximately 32,187 kilometres (12,874 kilometres a year). This is in line with the average annual mileage per car in England of 12,713 kilometres per year (RAC Foundation, 2017), but they are concerned about the amount of travel that they do by car. They have no bicycles. They are seeking to relocate to York in the next year to be closer to family, and aim to find a house in a suburb where activities are walkable, to reduce their reliance on the car and its associated costs.

References

Anon (2014) Vulcan Foundry, available at http://www.steamindex.com/manlocos/vulcan.htm, accessed 22 June 2017.

Arctic Flavours (2016) Finnish forests and bogs supply an abundant natural crop of berries, available at http://www.arctic-flavours.fi/en/info/berries/, accessed 20 September 2016.

Autoevolution (2017) Renault Scenic specifications, available at http://www.autoevolution.com/cars/renault-scenic-2003.html#aeng_renault-scenic-2003-16-16v, accessed 15 March 2017.

BusinessOulu (n.d.) *Oulu is growing and developing: We see to that*, available at https://www.businessoulu.com/en, accessed 4 May 2017.

City of Helsinki (n.d.) *Leisure in Oulu*, available at http://www.infopankki.fi/en/oulu/life-in-oulu/leisure-in-oulu#3, accessed 22 June 2017.

City of Oulu (n.d.) *Housing and environment*, available at https://www.ouka.fi/oulu/english/city-planning, accessed 22 June 2017.

Citypopulation (2016) *Newton-le-Willows (Merseyside in North West England)*, available at http://www.citypopulation.de/php/uk-england-northwestengland.php?cityid=E35001225, accessed 22 June 2017.

Climate-Data (n.d.) *Climate: Eschenlohe*, available at https://en.climate-data.org/location/504776/, accessed 22 June 2017.

European Space Agency (2017) *Flood-Europe, August 2005*, available at https://earth.esa.int/web/earth-watching/natural-disasters/floods/content/-/asset_publisher/zaoP2lUloYKv/content/flood-europe-august-2005, accessed 22 June 2017.

Heinonen V and Peltonen M (eds) (2013) *Finnish Consumption. An Emerging Consumer Society Between East and West*. Helsinki: Finnish Literature Society.

Kaden B (2008) *Neue Brücke in Eschenlohe*, available at https://ssl.panoramio.com/photo/15559689, accessed 22 June 2017.

Leigh H, Fonblanque A and Forster J (1830) The opening of the Liverpool and Manchester Railway – Fatal accident to Mr. Huskisson, *Examiner*, Sept 19 (1181), p. 603.

Lista A-L (2015) How Finland is fighting the sedentary lifestyle. WEST Welfare Society Territory. Available at http://www.west-info.eu/how-finland-is-fighting-the-sedentary-lifestyle/, accessed 19 September 2016.

Official Statistics of Finland (2015) Population structure, available at http://www.stat.fi/tup/suoluk/suoluk_vaesto_en.html, accessed 19 September 2016.

Official Statistics of Finland (2016) *Housing*, available at http://www.stat.fi/tup/suoluk/suoluk_asuminen_en.html, accessed 19 September 2016.

OKM (2014) OKM Education at a glance: Long study periods in Finland, Ministry of Education and Culture, available at http://www.minedu.fi/OPM/Tiedotteet/2014/09/EAG2014.html?lang=en, accessed 19 September 2016.

Pirinen P, Simola H, Aalto J, Kaukoranta J-P, Karlsson P and Ruuhela R (2012) Tilastoja Suomen ilmastosta 1981–2010. Raportteja 2012:1, Helsinki: Finnish Meterological Institute.

RAC Foundation (2017) Mobility, available at http://www.racfoundation.org/motoring-faqs/mobility#a24, accessed 15 March 2017.

Reher J (2013) *A healthy diet costs $2,000 a year more than an unhealthy one for average family of four: Harvard study*. National post December 6, 2013, available at http://news.nationalpost.com/health/a-healthy-diet-costs-2000-a-year-more-than-an-unhealthy-one-for-average-family-of-four-harvard-study?__lsa=b5ef-f3e7, accessed 20 September 2016.

Rolt L T C (1950) *The Inland Waterways of England*, London: George Allen and Unwin Ltd.

St. Helens Council (2017) Stepping through history: Newton Heritage Trail to launch on March 25, available at https://www.sthelens.gov.uk/news/2017/march/14/stepping-through-history-newton-heritage-trail-to-launch-on-march-25/, accessed 22 June 2017.

StatisticsTimes (2016) List of countries by population density, available at http://statisticstimes.com/population/countries-by-population-density.php, accessed 3 May 2017.

Swanson A (2016) *Ice cycles: The northerly world cities leading the winter bicycle revolution*, available at https://www.theguardian.com/cities/2016/feb/12/ice-cycles-northerly-world-cities-winter-bicycle-revolution, accessed 20 June 2017.

Trading economics (2016) Finland imports, available at http://www.tradingeconomics.com/finland/imports, accessed 20 September 2016.

Valtonen M, Laaksonen D E, Laukkanen J, Tolmunen T, Rauramaa R, Viinamäki H, Mursu J, Savonen K, Lakka T A, Niskanen L and Kauhanen J (2010) Sedentary lifestyle and emergence of hopelessness in middle-aged men. *European Journal of Cardiovascular Prevention and Rehabilitation*, 17(5), pp. 524–529, doi:10.1097/HJR.0b013e328337cced.

Vepsäläinen M, Strandell A and Pitkänen K (2015) Muuttuvan vapaa-ajan asumisen hallinnan haasteet kunnissa. *Yhdyskunta suunnittelu. The Finnish Journal of Urban Studies*, 53, available at http://www.yss.fi/journal/muuttuvan-vapaa-ajan-asumisen-hallinnan-haasteet-kunnissa/, accessed September 19 2016.

Vinnari M, Mustonen P and Räsänen P (2010) Tracking down trends in non-meat consumption in Finnish households, 1966–2006, *British Food Journal*, 112(8), pp. 836–852.

WHO (2015) Using price policies to promote healthier diets. WHO European Regional Office, available at http://www.euro.who.int/__data/assets/pdf_file/0008/273662/Using-price-policies-to-promote-healthier-diets.pdf, accessed 19 September 2016.

Worldometers (2017) Population: Europe, available at http://www.worldometers.info/population/europe/, accessed 3 May 2017.

Yr (2017) Weather Statistics for Newton-le-Willows (England, United Kingdom), available at http://www.yr.no/place/United_Kingdom/England/Newton-le-Willows/statistics.html, accessed 22 June 2017.

5
NORTH AMERICA

Fabricio Chicca, Silvio Marco Costantini, Abbie McKoy, Ana Paula Pagotto and Brenda Vale

Background

Brenda Vale

The continent of North America is made up of 23 countries. Two are very large—the United States and Canada—and make up 79 per cent of the land area of the continent (Countries of the World, 2017), while the balance comprises a number of smaller and very small ones, the latter mostly island countries in the Caribbean. North America ranks third in terms of continental size, behind Asia and Africa, occupying 16.5 per cent of the land area of the planet, but is the fourth largest continent when it comes to population as Europe has almost 80 per cent more people (What Are the 7 Continents, 2017). The largest city is Mexico City, with a population of just over 21 million in the greater region and almost 9 million in the city (World Population review, 2017a). Stretching from the Arctic Circle towards the equator, where the continent's southernmost country of Panama joins the land mass of South America, and from very small islands to a continental land mass, climates are very varied. Brief descriptions of the climates of the places where the stories are located are given below, together with more local detail.

When it comes to ecological footprint (EF), the big countries (the United States and Canada) have big EFs. In 2011, these were 7.1 and 6.7 global hectares respectively, both falling among the ten countries with the highest global footprints. In 2011, the average EF of Mexico was just over the world average of 2.7 global hectares, at 3.0 global hectares per Mexican. North American countries falling below the fair share EF of 1.8 global hectares included Guatemala, Honduras and Jamaica, with Haiti having the fourth lowest EF in the world, at under 1.0 global hectare per person (WWF, 2012).

Toronto, Canada

Originally established in 1793, the settlement grew until in 1834 it became the city of Toronto (City of Toronto, 2017). Today the Greater Toronto Area (GTA) is the largest metropolitan area in Canada and fourth largest in North America (City of Toronto, 2017). In 2015, the population of the GTA was over 6.6 million, just over 18 per cent of all Canadians, and the city itself was home to 2.8 million (Country Digest, n.d.). Toronto is a multi-cultural city as in 2011, 46 per cent of the GTA population were immigrants (Statistics Canada, 2016), and is the financial and business capital of Canada (City of Toronto, 2017).

Situated on the northwest shore of Lake Ontario, the GTA covers a gently sloping plateau cut across by many rivers that drain into the lake (City of Toronto, 2017). The climate is cold-temperate and it rains every month, the driest month being February, with 51 millimetres of rainfall, and the wettest August, with 81 millimetres. Temperatures range from −5.3°C in January (lowest monthly average) to 21.5°C in July (highest monthly average) (Climate-Data, n.d.).

An old study measured the EF of Toronto as 7.6 global hectares/person, which was just below the Canadian average of 7.7 global hectares/person (Onisto et al., 1998). Since then, the city has initiated a programme to make it "the most sustainable city in North America", but mention is no longer made of the EF (City of Toronto, 2017.

Havana, Cuba

Before Fidel Castro came to power in 1959, Havana was a US tourist destination for its climate, harbour setting and historical colonial architecture. In 1989, after 32 years under Castro, the tourist facilities were no longer up to North American standards (Halperin, 1992). Havana has changed, and for Halperin, for the worse, but other people feel differently. "[Havana] … is in a way our ideal of the dense mixed city: organised, organic and chaotic it shows up our well-serviced copies for the pale imitations they are" (Rattenbury, 2006). The story below illustrates how this works.

For those interested in sustainability, Havana has become strongly associated with urban agriculture, which became necessary after the fall of the Soviet Bloc and its detrimental impact on Cuba's economy. In Havana, the local government encouraged the conversion of parks, sports fields and demolition sites into vegetable gardens, and food production in private gardens soon followed (Premat, 2009). This urban agriculture has since been linked to making a city more sustainable (Cruz and Medina, 2003: 195; Gold, 2014).

Havana has a population of over 2.1 million (World Population Review, 2017b). The climate is tropical. The hottest months are July, August and September, with daily maximum temperatures averaging just below 30°C, and minimums around 24°C. Summer is when most rain occurs. August and September are also the hurricane season. The coolest months are January and December, with daily maximums of around 25°C and minimums around 16°C (World Weather and Climate Information, 2016).

St. Tammany Parish, Louisiana, United States

The state of Louisiana is divided into parishes. The rural parish of St. Tammany lies on the northern shores of Lake Pontchartrain, with New Orleans to the south of the lake. St. Tammany was established in its current form in 1810. Before this time, it was "owned" by Spain, France, England and then Spain for a second time (St. Tammany Parish Government, 2017), not forgetting it was once home to Native Americans, including the Choctaw tribe (Gregory, 2015; Bushnell, 2006). The parish is now (2015/2016) home to 233,740 people, of which 0.5 per cent are Native American (US Census Bureau, n.d.).

The climate is warm with no snow but more rain annually than the US average (1,588 millimetres and 996 millimetres, respectively). July is the hottest month, with an average temperature of 33°C, and January the coldest, with an average of 4°C (Sperling's BestPlaces, 2017). The area is a high hurricane risk zone, with 68 events since 1930 (Homefacts, 2017).

Celebration, Florida, United States

This is the only story set in a new town. When the second Disney theme park was established near Orlando in Florida, the Disney Corporation bought up land to create a buffer zone around it (Stringham et al., 2010). In 1994, some 26 square kilometres of this buffer zone became the settlement of Celebration, with a planned population of 10,000. Based on the ideas of neo-traditionalism which were derived from New Urbanism (Frantz and Collins, 1999: 43), Celebration was conceived as "a living laboratory for the American town" (Frantz and Collins, 1999: 47). Residents were placed in a series of villages around the town centre and streets were designed to be walkable. Everything was to be available within the confines of the town unless you happened to work outside it (Ross, 1999: 294).

Celebration has a sub-tropical climate with no snow, apart from the Christmas season artificial snow falling on the plastic ice rink in the town centre (Pilkington, 2010). Temperatures range from 10°C–21°C in the coldest months to 21°C–32°C in the warmest, with summer being the wettest time of year (City-Data, 2017).

Hunter and Sadie: Living downtown: Canada

Researcher and writer: Marco Costantini

The condominium

Hunter and Sadie live in Toronto, in the province of Ontario. Temperatures range from −22°C in winter to +16°C in summer. The couple rent a unit in a large condo (condominium) complex very near the downtown core of the city. The condo is a new development, and they are the first residents of their particular unit. Hunter works as a software engineer downtown, and Sadie as an administrative assistant in a nearby city. They chose to live downtown because of the proximity to the city attractions.

The condo is very small by Canadian standards, at just over 60 square metres in area. At the time of the story, their rent payment was C$1,600 per month. The unit has one bedroom, an open space which comprises the kitchen and living area, and a den, which Hunter uses as an office. There is a closet next to the entrance for laundry appliances, and on the other side of the entrance another closet for hanging coats and storing shoes and boots. At the rear of the condo is a balcony shared with the adjacent condo, but separated from it by a partition. The balcony overlooks the courtyard, which is surrounded on three sides by the condominium buildings. The condominium complex has several shared resources. These include a basement recreation area which is staffed during the day. There is also a large swimming pool and hot-tub, a gym, a squash court, a billiards table, a table-tennis table and a sitting area for relaxing.

There are several appliances and electronic items in the home. Their kitchen has an electric stove and oven, a microwave, a dishwasher and a refrigerator/freezer. Their laundry area has a front-loading washing machine, with a front-loading dryer above it. The office has a computer system including three laptop computers. The living room has a 47-inch flat-screen television and many video-game consoles, as video-gaming is Hunter' primary hobby. The bedroom has a queen-size bed.

Transport

In Canada, the car is the key component of family transportation, and this couple are no exception to that rule. They have one vehicle, a 9-year old sports car which is not very comfortable, and does not have much storage space. In a sense, this car does not suit their needs, but it was loaned to them by other family members when they needed a car. Sadie takes the car to get to work and Hunter uses public transport. Sadie takes approximately 30 minutes for the 35-kilometre trip from home, using toll-free highways. Approximately C$60 worth of fuel covers the trip to work and back for the whole week (Monday to Friday). Hunter walks approximately 10 minutes to get to the most convenient street-car stop. Once there, he takes a street-car for approximately 1 kilometre, after which he walks for a further 5 minutes to the downtown building where he works. Toronto has a large underground network of shopping centres called "The Path". In the cold winter, it is desirable to walk through this as opposed to walking outside, and from the street-car to his workplace, Hunter walks through the Path. Each month, Hunter purchases a monthly public transportation pass that costs C$138. This allows him to use any mode of transportation provided by the Toronto Transit Commission. The couple visit friends and family at the weekends. This almost always requires the car, and they could travel as far as approximately 112 kilometres. The average visit was to Hunter's family approximately 25 kilometres from home.

Food

The couple eat an even distribution of home-prepared and store-bought meals. Hunter usually skips breakfast but Sadie normally has some store-bought biscuits or toast with tea. She usually makes her lunch at home and takes it to work. This is often food left over from dinner the night before, otherwise it is food bought from the grocery store specifically for her lunch. Hunter buys lunch from a fast-food restaurant every day, and this might be stuffed pitas, kebabs, Japanese food, Indian food, Chinese food or sandwiches. It is convenient for him to do this, as there are many restaurants in the Path below the building where he works. Without exception, all of his lunches included meat in the week described. The usual cost is approximately C$10, but sometimes he spends more. This was usually the case when he joins others for lunch and cannot control the venue. The couple like to eat fast food a few times each week, which might be hamburgers and french fries from a drive-through fast-food restaurant, or roast chicken with french fries. However, this is not an easy thing to do as there are no restaurants within a reasonable walking distance of the condo and driving downtown to get food is challenging. As a result of living in the condo, the couple have found themselves making more meals at home than they did before. There is a grocery store very close to their home, and Hunter often buys food there for the day on his way home from work. This is also costly, as they felt they were not adept at cost-effective cooking. They often cooked just one meal, which might be roast beef and mashed potatoes; sandwiches with beef, lettuce and tomato; chicken and vegetable stir-fry; or spaghetti bolognese, although leftovers became next day's lunch for Sadie. Hunter likes to snack on chocolate biscuits in the evening while Sadie prefers fruits, such as plums and nectarines.

Energy

Electricity for the condo is provided by a company called Toronto Hydro. Electricity is colloquially called "hydro" because much of Toronto's energy comes from the hydroelectric generators at Niagara Falls. Relative to large households, the cost of electricity for the condo is small. For a very moderate month in Toronto (May,) the electricity invoice was C$53.92, and there is a substantial charge for just delivering the electricity of C$28.56 (53 per cent of the total). Electricity usage is categorised as "on-peak", "off-peak" and "mid-peak". The C$25.36 of electricity consumed breaks down into C$5.82 on-peak (just over 45 kilowatt hours), C$2.49 mid-peak (nearly 23 kilowatt hours) and C$13.54 off-peak (88 kilowatt hours). This indicates that most electricity consumption happened during off-peak hours. These start at 7 pm and go into the night. This manner of usage is primarily due to the fact that both Hunter and Sadie are at work during the day, and only arrive home around 6.30 pm. In the week described, the couple used appliances whenever it was convenient for them and did not set out to maximise their off-peak and mid-peak use.

The condo has central heating and air conditioning. While the couple could set the desired air temperature using an electronic thermostat, they could not control whether the heating and air conditioning systems were on or off. These systems were managed by the condominium staff according to the time of year. The charge for this service is included in the rent.

Silvio and Veronica: Watching television in the street: Cuba
Researcher and writer: Fabricio Chicca

The family

Silvio and Veronica, both 32 years old, live in Old Havana (*Havana Vieja*) with their son Ricardo who is 12 years old. They live in an old 1930s apartment building with high ceilings. The building is organised around a long access corridor. Four apartments have their entrances from the corridor, which leads to an internal courtyard around which eight other apartments are organised. One of these is the flat where Silvio's parents live. When Silvio and Veronica were officially married in 2005, with the help of his sister who lives in Spain, and thanks to the high ceiling, they managed to build an intermediate floor within the parents' apartment. It took them almost 3 years to finish the renovation. In order to be accepted as a new residence, an independent access had to be created. As a result, the main door of the flat opens into a tiny hall, with two doors. One of the doors leads to Silvio's parents' apartment on the lower level, and the second one to the new upper-level apartment, via a flight of very narrow stairs.

The apartment

Their flat on the new intermediate floor is around 35 square metres, divided into four spaces. Facing the street is a combined living room and kitchen, which contains a small two-seater sofa, a dining table, two bookshelves, a 24-inch television, a stereo and a fan. There is a small corridor off of which is an alcove where Ricardo sleeps, and at the end a bedroom and toilet. The family stores their belongings in a wardrobe in the middle of the corridor. Because the apartment has been inserted into an existing one, it has a very low ceiling height of around 2.1 metres. Their bedroom has a small window on to the building's main access corridor and their living room has two windows, which are in fact the top half of the windows that light the living room of the parents' apartment below.

Electricity

Silvio and Veronica pay around CUP72 a month for electricity, which equates to using 180 kilowatt hours a month. The price for electricity increases according to its consumption and the family have made an effort to reduce their consumption from CUP108 a month to CUP72. They watch less television, and open the refrigerator and turn the lights on less often. Veronica thinks they cannot reduce their consumption any further.

Work

Silvio is an engineer and Veronica works at the Havana port. Silvio also has a second job, which provides the highest proportion of the family income. After work, he drives his cousin's tourist cab. His cousin has the official cab license and he drives the cab from 4 am to 4 pm. Silvio takes over from 4 pm to 8 pm, and Silvio's brother drives the cab for the rest of the night. The cab thus runs 24 hours. Silvio did not explain the detailed financial arrangements, but the income from the taxi is at least five times the income from his day job.

Daily routine

The family's day starts around 7 am when the three of them take turns to go daily to the bakery 50 metres from their house. Breakfast is always coffee, bread with butter and guava marmalade and fruit. After breakfast, the family leave the house together around 8 am. Silvio takes a bus, or a communal taxi. He first walks six blocks out of the old city, and then takes the bus. He occasionally takes the taxi when he is late or if the bus is too busy. Veronica needs to walk four blocks to the port in Old Havana. Ricardo walks to his school in the middle of the old city. They each carry a little container with food for lunch, which normally consists of leftovers from the previous dinner. Ricardo also has a sandwich or fruit. He is the first one to return home, around 4 pm. He first hangs out with his grandparents, where he finishes his homework, and then goes to play in the street with his friends. Next year, his routine may change, as he will be sent to improve his baseball skills. For the past 3 years, he has been nominated the best baseball player in his school.

Veronica comes back from work around 4:30 pm. She then prepares the dinner, checks Ricardo's homework and takes care of the housekeeping and laundry, although she normally manages to have some spare time to sit in the street outside to chat with the neighbours. The family have dinner after 8 pm when Silvio gets back from his second job. Old Havana has a strong sense of community and family relationships are strong inside this neighbourhood. Lately, the area has been affected by the flow of tourists, and their potential for generating extra revenue. Silvio's family benefits from this as his second job mostly involves tourists and the majority of these use a special currency (CUC). One CUC is equal to one Euro, which is worth 25 Cuban Pesos. A taxi ride from Havana airport to town costs around CUC25. This needs to be compared with the monthly salary for a university lecturer of CUC45. Although what Silvio earns from driving his cousin's taxi greatly exceeds that from his normal day's work, he keeps his work as he considers the arrangement with his cousin to be temporary because his cousin is the license holder. On the other hand, the current jobs of Silvio and Veronica, with their much lower pay, entitle them to receive benefits from the government. The most important one, which the family greatly appreciates, is food subsidies. The food they are allowed to buy from the government is very cheap. These basic supplies are distributed through a ration book known as the *Libreta de Abastecimiento*.

Food

Although the family claim that the food from the government does not last the entire month, they also recognise the benefits of having such a system. According to the family, the amount of food available varies according to the season and availability of goods. When a product is not available it is simply not issued, and in most cases, the family receives a double amount the next month. The main products in the ration book are rice and beans. Meat is included in a different list, and variations in the meat distribution are very common, as the supply of meat, pork, chicken and fish is not guaranteed. Because the food obtained through the ration book is not enough for the family, further supplies have to be purchased on the market. The list of subsidised products includes beans, rice, olive or soya oil (depending on availability), refined and brown sugar, salt, coffee, shower and cleaning soaps, toothpaste, jam, eggs and milk. The amount of food changes according to the size and age of the family members. Each family receives its own booklet, which is only valid in the food centre (*bodega*) nearest their house.

During the time of this story, Veronica went to their *bodega* to buy the products to which they were entitled. On that occasion, all products were available and in the correct amount, even chicken. The family rely on food from the official *bodega*, the local bakery and the vegetable market near their home. They also often buy meat and pork on the black market, which is sold from a building beside their home. The meat black market in Old Havana is very common, and as long as you have money, meat is fairly available. The prices on the black market are somewhere between those of the official *bodega* and the dollar market. The dollar market is found in the old shops that used to operate in American dollars and which are now mostly focused on tourists. They are still around, but most customers are now tourists and local Cubans who have managed to increase their income.

The main family meals are based on rice and beans with the addition of some meat. In the absence of meat, roots such as taro and sweet potatoes are added. One night, Veronica prepared a chicken stew that lasted two days. On the other days, they had rice and beans and taro, pasta with tomato sauce, rice and beans with a local sausage and, on Sunday, rice and beans and fried pork. Veronica cooks every night and the leftovers are carefully planned to provide lunches for the next day.

Social events

On Saturday morning, the family went to a meeting with the other residents in their neighbourhood block. These meetings happen often and are part of the social and political arrangement. The residents have the opportunity to express their concerns about their local area. The meetings are called Committees for the Defence of the Revolution (CDR) or *Comités de Defensa de la Revolución*[1]. The meetings bring together the entire block, and are an opportunity not only for social but also political engagement. During the meeting, Silvio was proposed as the block representative candidate for the next municipality election, but he declined. After the meeting, there is an informal gathering. Families bring their own chairs and sit in the narrow streets for a chat and a cigar. The meeting finished around 5 pm. While Veronica stayed and socialised with other families, Silvio left for his second job. He returned at the usual time, and the entire family went for a walk along the Havana waterfront. They had pizza from one of the hundreds of small places that serve pizzas on the go. Pizzas only cost 20 Pesos and are an inexpensive way to eat out.

What happened after the CDR meeting is what happens most nights. Families put their chairs outside their houses, and the street is taken over by people. All doors are open, and families move freely between houses, while the kids run around. This is part of the family's routine which Veronica and Ricardo enjoy. When Silvio comes back home, the family watch television together. Two types of programme are very popular with the family and the other families around. One is Brazilian or Mexican soap operas, and other the baseball games. If it is too hot to watch television when Silvio comes home from work, they sit outside and watch his parents' television, which is put on the window sill.

Grant and Anthea: After the hurricane: United States

Researcher and writer: Abbie McKoy

Hurricane Katrina

St. Tammany Parish has never been the same since Hurricane Katrina hit in August 2005. For many decades, this rural Louisiana area was completely remote and isolated. It has recently experienced significant change and considerable growth, with many people moving there from New Orleans and surrounding suburbs. Yet, despite pockets of growth, many of the rural areas in this parish maintain their isolation (USDA, 2016). Members of this community typically start conversations with strangers by referencing their Hurricane Katrina story. This major storm changed the lives of everyone in the area, and this reality was particularly apparent during the time spent with the couple described here.

The dwelling

Grant and Anthea own nearly 10 hectares of land, and the couple live in a beautiful 288-square metre home constructed of brick and plasterboard, with white vinyl siding and a concrete foundation. The roof is made of asphalt shingles. The home is slightly raised off the ground in order to allow airflow under the house and to help if flooding occurs.

A 56-square metre apartment sits over the garage and is made of the same materials. A large barn located toward the front of the property has a metal roof and is of commercial-grade steel construction. Several small structures of recycled wood are spread across the property. The family's home and all of the other structures were designed and built by Grant. The couple are wealthy as a result of his successful company, which serves customers all across Louisiana and beyond.

Life before Hurricane Katrina

In order to explain their current way of life, the couple first described their life before Katrina. As Anthea looked out her kitchen window, she reminisced about how that storm had changed everything. Although the hurricane had occurred more than a decade before, her face still showed apprehension and even terror when she recalled what happened. During the storm, she and Grant hid in the bathroom with their two dogs. As their entire house shook violently in the wind, they felt sure they would perish. Trees fell through the roof, windows broke, and all they could do was to hold each other tight, praying that this horrific storm would soon come to an end. Hours passed with no relief, leaving them to wonder what the outcome would be.

Finally, the winds subsided, but they waited a while longer, afraid to face what might lie beyond the bathroom door. When they finally ventured out, they were shocked by what they found. Trees had caved in the roof, and their entire home lacked windows or shape. As they made their way outside, they saw their land in a new light. The trees which once isolated them from the world lay broken across the lawn and driveway, blocking any exit in their vehicles. All their crops were destroyed. Their well could not be accessed, as debris blocked its entrance. It soon became clear that their challenges had just begun. They didn't have food or water stored. They lived far from their neighbours and were uncertain whether anyone would even find them.

Weeks passed and no one came to offer assistance. Their phones had no connection. They were alone and realised they would need to be self-sufficient in order to survive. So they worked tirelessly on sawing the trees into pieces to clear a way to the main road. Once they had cleared their driveway, they were able to assist neighbours. With trees in the community having fallen across the roads, weeks turned into months with no outside assistance. Things were far from "normal" at this point. Credit cards and cash were both useless, so they bartered, exchanging medical supplies and other items for food.

This experience changed the couple forever, causing them to make several life-altering decisions. Basically, they became survivalists. This meant creating a second well on the property, and they are now in the process of constructing a third. They also decided to create three gardens on the property. These were purposely placed in different locations to make the odds better that one garden would survive even if the others were destroyed by a storm or flood. They acquired chickens, starting with three and now up to twelve. They began pickling, dehydrating and canning food so they could keep it for years. They began making their own wine, primarily for bartering. They collected guns for hunting. They stored blankets and other necessary supplies in different areas across their property.

The couple have become pillars of their community, networking with others with the understanding that they can help each other in times of need. I happened to visit at another historic time for this area. A flood had hit St. Tammany Parish, causing hundreds of homes to be damaged (Adelson et al., 2016). Once again, many families were left without any assistance from government agencies or social services organisations, but this time the community came together in a much better way. Community members pitched in to donate clothes, food, diapers, cleaning supplies and their time to help people rebuild their lives. The couple were central to this community-based effort, providing many items crucial to those impacted and generously giving their time to the cause by assisting at a local storefront that had been donated as a resource centre after the flood.

Self-sufficiency and food

As a result of surviving these natural disasters, the couple are much more self-reliant and much less dependent on the outside world. Each morning they collect eggs from their chicken coop for breakfast. Usually, they eat one or two eggs, and they always have more eggs than they are able to eat. They often pickle the small ones and share the larger ones with their neighbours, sometimes bartering for other produce. They also harvest fresh corn and make grits, a southern dish created from cornmeal. They accompany this with fresh juice made of beets, kale and blueberries. They have oats on some mornings, and these are supplied through an exchange with one of their neighbours.

For lunch, the couple often eat salad topped with local fish or shrimp. Grant purchases seafood from a local market, gets it by fishing in the surrounding streams or barters for it through trading with neighbours. Lunch is often accompanied by more fresh juice, which Anthea feels is an important part of her health regimen. For dinner, they typically eat meat from the local market, along with potatoes and fresh salad harvested from their garden. The meats include chicken, wild turkey, grass-fed beef, local fish and venison. They are able to catch wild turkey and deer in the woods surrounding their home, though sometimes they still have to buy these from the local supermarket. They snack once or twice a day, often eating nuts, dried fruit, pickled eggs or smoked meat. They rarely eat out, on average about once a week, but sometimes they get a salad at their local gym after a workout.

Transport

The couple have two personal vehicles—an old largish saloon car and a modern luxury sports car. They also have two company cars—a diesel pick-up truck and an SUV. In addition to these, they have a diesel "side-by-side" farm vehicle that they use on the property every day in order to go quickly from one side of the property to the other. They also have a quad bike, two bicycles and a boat. They leave their home 5 days a week in order to use the local gym, get physical therapy for Anthea or obtain supplies from various stores in town. Their commute averages about 92 kilometres a day.

Grant also goes to work about two to four times a month, but the majority of his work is done remotely at home using the internet and a phone to manage his staff. Some of his employees have a larger carbon footprint, with several commuting each day all the way from parts of New Orleans, a roundtrip commute of 128–178 kilometres. In order to be more sustainable and economical, the company replaced all of its trucks with small cars which use only 5.9 litres per 100 kilometres. Their area used to not have phone service, and even now there is only one company providing this. Without the internet and a reliable phone service, it would have been impossible to run the company remotely.

Energy

New costs have been applied to utilities in St. Tammany Parish, and it is likely some of this extra expense was caused by the aftermath of Katrina and other storms. In order to conserve energy, the couple do not use central heating. Instead, they use a wood-burning heater, which produces enough heat to warm the entire house. Local electrical power comes from coal and nuclear energy sources in the neighbouring state of Mississippi. This electrical service is provided through a cooperative that provides power through local community negotiation rather than from a public utilities electric company. Their energy usage averages 1,037 kilowatt hours per month, which is a little lower than the average in the area of 1,184 kilowatt hours.

Because of the wells on their property, the couple have no water bill. They also have no sewage bill, as they have a septic tank on site. Due to the severe heat in the summer, they typically use air conditioning from April to October. They have two separate air-conditioning units, one in the master bedroom and the other providing cool air to the entire house. They have solar outdoor lights. They also have a solar generator which is able to support the apartment, a whole-house generator for the main home and three gas generators (one of which is used to power the barn).

The family have many appliances, including a washer and dryer, stove, crock pot, microwave, food dehydrator, commercial kitchen, coffee maker, rotisserie, pressure cooker, smoker, vacuum packer, canning pot, Ninja blender, Vitamix, gas grill, Brazilian disc for cooking over the fire, George Foreman grill, meat grinder, sausage stuffer, meat saw, meat slicer and commercial knives. They have a pool where kids and families can come to bathe in case of emergency. They also have a small sawmill to turn trees into lumber, and they have already used some of the wood to make the barn, fence and chicken coop.

Craig and Shirley: Life with Disney: United States

Researchers and writers: Fabricio Chicca and Ana Paula Pagotto

The family and their house

The family living in Celebration, Florida are made up of the parents (Craig, 44, and Shirley, 51) and a girl and boy of 4 and 8 years old, respectively. Their three-storey house is approximately 210 square metres in area. It was built in 2001, using a timber structure with a stucco finish to the exterior walls and a concrete floor slab. Inside there are three bedrooms, three bathrooms, three living rooms and a garage for two cars. It has two space conditioning units, so the whole house can be heated and cooled as necessary. As part of its furnishings and fittings, it contains five televisions (two televisions are 33 inches, two are 42 inches and one is 22 inches), a refrigerator, stove and oven, microwave oven, dishwasher, washing machine, dryer, electric heater and two computers.

Food and routine

In general, at breakfast, the family consumes 1 litre of milk with chocolate powder for the children and 250 millilitres of coffee and around 500 millilitres of orange juice for the adults. Between them they eat two slices of toast and six rolls, which are bought canned and frozen and then baked every morning. These are normally eaten with butter and jam or cream cheese.

The children have their lunch at school. However, the son, who now attends a public school, does not like the food served in its canteen, and as a result, during the day he only eats a piece of bread with olive oil and an ice cream. When he arrives home he is starving and has a large meal. He normally has meat and potatoes with rice or pasta bolognese. These meals are accompanied by two glasses of chocolate milk. During the week described, he ate no other vegetables or fruit or natural fruit juice. In contrast, his sister has a more balanced diet. She attends a different school. For the morning snack she has fruit, carrots, celery and peanut butter. Her lunch, which she takes from home, is leftovers from dinner the previous night. When she gets home around 4 pm, she drinks some milk with a cookie and occasionally has fruit. At night, she either has a light meal or cereal with milk. According to the family, she has the best diet of them all. The family do not sit down to eat together, but meals are taken according to each individual's need and appetite. The children have been brought up to grab something to eat whenever they are hungry. The girl will eat any food available in the refrigerator. The boy loves the simple meal of meat, potatoes and rice, so this is always available. He fries his own potatoes using an air fryer. In the week described, on occasions when these foods had run, out they stopped at a fast-food outlet for chicken nuggets and fries.

During the period described, Shirley was on a diet. Before setting out for work she used her phone to order a fruit smoothie and a quesadilla. She grabs this meal and eats it before getting to work. This is the only substantial meal she has during the day, as her work schedule is very tight, leaving little time for eating. She says her only fruit intake comes from the smoothies. When she gets home with the children, she has a cup of coffee and a slice of bread. Before bed, she has cereal, or a glass of chocolate milk. Craig has a different routine. After breakfast, he goes to university and from there, around lunch time, he goes to his work. Before work, he goes with his workmates to lunch, patronising one of several local fast-food options. On Monday, he ate Mexican fast food; Tuesday, he had a Chinese meal; Wednesday, he tried a Brazilian buffet restaurant; on Thursday, an American fast-food outlet; and the Chinese once again on Friday. Every meal was taken with fizzy drink.

At night, he has his main meal, which is essentially the same meal as his son (meat and potatoes with rice), together with a salad and a beer.

As well as not eating together, each family member has his or her own routine. Both children have a tablet, on which they spend most of their time at night before and after dinner. Craig likes to spend a couple of hours watching television in the evening, while Shirley catches up with friends through social media. At night, before bed, the children again have milk with chocolate.

On Saturday, as Craig was working, the rest of the family went to a fast-food outlet for lunch. After work, Craig bought his lunch of chicken with salad and potatoes at the grocery shop. For dinner, the family had a pizza, which they ate together at home. They go together to the supermarket every Sunday after church for the bulky groceries. Shirley occasionally visits the supermarket during the week for milk, bread and small items, as she works inside a mall in front of the supermarket. During the week described, the following grocery items were bought:

- 12 litres of milk
- 2 litres of juice
- 4 litres of soft drinks (bottle) + 9 cans
- 12 beers (cans)
- 1 bottle of wine
- 38 litres of water
- 30 yogurts
- 1 pack of toast bread
- 3 packs bread cans
- ½ pound (227 grams) margarine
- ½ pound (227 grams) butter
- ¼ pound (113 grams) jam
- 250 grams of coffee
- 2 kilograms of chocolate powder
- 500 grams of refined sugar
- 250 grams of beans
- 500 grams of rice
- ½ litre of soy oil
- 1 pack of frozen potatoes (3 kilograms)
- 6 eggs
- 6 frozen burgers
- 4 bananas
- 4 apples
- 250 grams of pasta
- ½ cookie package
- 1 bag of chips (crisps to English readers)
- 1.2 litres of cleaning products
- 2 rolls of toilet paper and 1 roll of paper towel.

Transport

The family have two cars. The family car normally used to take the children to school is a conventional smallish car with a 1.4-litre engine (9 litres per 100 kilometres). Craig drives a hybrid car (4.6 litres per 100 kilometres). The family also have two electric scooters and three bicycles. They feel public transport is not an option for them where they live.

Shirley is responsible for taking the children to and from school using the conventional car. However, her son at 8 is now capable of going to school by himself, and cycled there two out of the five days in the week being described. After dropping the children off, Shirley goes to work in the car. Her entire daily journey is around 31 kilometres. Craig, however, drives more during the week, going from home to university and then work, a total round trip of around 54 kilometres a day in the hybrid.

Energy

The family have recently moved to this house and energy consumption in April averaged 35 kilowatt hours per day. However, the consumption is predicted to double during the hotter months of the year.

Leisure

Celebration is a very small community, divided into four villages (North, South, East and West). Each has its own leisure space or spaces and the entire Celebration complex has nine such spaces, confusingly also called villages, which anyone in the community can use. All the leisure spaces are similar with swimming pools, a pitch for American football and soccer, basketball courts, tennis courts, fitness centres, ballrooms, business and meeting rooms, dance rooms etc. The family's social activities revolve around these places. After their homework, around 5:30 pm, the children go to the village nearest their house for activities such as football (three times a week) and swimming (twice a week). The family pay an annual fee to use the villages and a top-up fee for each specific activity.

Every Friday after work, the whole family goes to a friend's house, or receives friends at home for a catch up, usually accompanied by drinks and snacks.

During the weekends, the family's activities are also related to the villages. On Saturday mornings, after his breakfast, Craig goes to work, as he works at the weekend as well and the rest of the family goes to the villages. On the week in question, this was for a football game, and was followed by lunch at a fast-food chain. After lunch and a short rest, Shirley and the children went back to the village to enjoy the swimming pool for a couple of hours, meeting friends there, again demonstrating the importance of the villages for their social life. Later in the day, they went for a bicycle ride. They played games in the park beside the village before coming home to have dinner with Craig.

On Sundays, the family have a slightly different schedule. In the morning, Craig goes to church alone, while the rest of family sleep until late. They get up around 11 am for a late breakfast and then go to the village. Craig is reunited with them around noon. They have their lunch in a fast-food outlet and then return to the village, where they spend the rest of the afternoon together. Around 5:30 pm Shirley goes to the church, coming back around 6:30 pm; she picks the family up and they all go to the supermarket for the weekly groceries and back home for dinner.

Note

1 Outside Cuba, the term CDR is considered to be a mass vigilance tool of the government. During this study, criticisms against the government were fairly common in private spaces, and moderately so in public areas. However, the idea of the CDR as a mass vigilance tool was dismissed by the locals.

References

Adelson J, Antonio Vargas R and Pagones S (2016) St. Tammany Parish president Pat Brister on flooding: 'I've never seen anything like it'. *The Advocate*, available at http://theadvocate.com/news/15166566-148/flood-updates-in-new-orleans-area-50-in-covington-rescued-overnight-as-flood-levels-quickly-rose, accessed 20 March 2016.

Bushnell D I (2006) The Choctaw of Bayou Lacomb, available at http://genealogytrails.com/lou/sttammany/choctaw.html, accessed 27 June 2017.

City of Toronto (2017) *Toronto History*, available at http://www1.toronto.ca/wps/portal/contentonly?vgnextoid=4284ba2ae8b1e310VgnVCM10000071d60f89RCRD, accessed 11 May 2017.

City-Data (2017) Celebration, Florida, available at http://www.city-data.com/city/Celebration-Florida.html, accessed 11 May 2017.

Climate-Data (n.d.) *Climate: Toronto*, available at https://en.climate-data.org/location/53/, accessed 19 June 2017.

Countries of the World (2017) List of countries in North America, available at https://www.countries-ofthe-world.com/countries-of-north-america.html, accessed 11 May 2017.

Country Digest (n.d.) *Toronto population 2017*, available at http://countrydigest.org/toronto-population/, accessed 19 June 2017.

Cruz M C and Medina R S (2003) *Agriculture in the City: A Key to Sustainability in Havana, Cuba*, Kingston, Jamaica: Ian Randle Publishers; Ottawa, Canada: International Development Research Centre.

Frantz D and Collins C (1999) *Celebration, U.S.A.: Living in Disney's Brave New Town*, New York: Henry Holt and Co.

Gold M (2014) Peasant, patriot, environmentalist: Sustainable development discourse in Havana, *Bulletin of Latin American Research*, 33(4), pp. 405–418.

Gregory H F (2015) Indians and folklife in the Florida parishes of Louisiana, available at http://www.louisianafolklife.org/LT/Virtual_Books/Fla_Parishes/book_florida_indians.html, accessed 27 June 2017.

Halperin M (1992) Return to Havana, *Society*, 29(6), pp. 53–62.

Homefacts (2017) Hurricane information for St. Tammany Parish, LA, available at http://www.homefacts.com/hurricanes/Louisiana/St.-Tammany-Parish.html, accessed 27 June 2017.

Onisto L J, Krause E and Wackernagel M (1998) *How Big Is Toronto's Ecological Footprint?* Centre for Sustainable Studies and the City of Toronto, Toronto.

Pilkington E (2010) How the Disney dream died in Celebration, *The Guardian*, available at https://www.theguardian.com/world/2010/dec/13/celebration-death-of-a-dream, accessed 11 May 2017.

Premat A (2009) State power, private plots and the greening of Havana's urban agriculture movement, *City and Society*, 21(1), pp. 28–57.

Rattenbury K (2006) Viva Havana: With developers poised to swoop down on Cuba, what will become of Havana's architecture? Kester Rattenbury says western models of redevelopment may destroy this extraordinary cultural marvel, *Building Design*, 21(1718), p.8.

Ross A (1999) *The Celebration Chronicles: Life, Liberty and the Pursuit of Property Values in Disney's New Town*, New York: Ballantine Books.

Sperling's Best Places (2017) St. *Tammany Parish, Louisiana: Climate*, available at http://www.bestplaces.net/climate/county/louisiana/st._tammany, accessed 27 June 2017.

St. Tammany Parish Government (2017) *History*, available at http://www.stpgov.org/i-want-to/learn-about/history, accessed 27 June 2017.

Statistics Canada (2016) *Immigration and ethnocultural diversity in Canada*, available at http://www12.statcan.gc.ca/nhs-enm/2011/as-sa/99-010-x/99-010-x2011001-eng.cfm, accessed 19 June 2017.

Stringham E P, Miller J K and Clark J R (2010) Internalizing externalities through private zoning: The case of Walt Disney Company's Celebration, Florida, *Journal of Regional Analysis and Policy*, 40(2), pp. 96–103.

United States Department of Agriculture (USDA) (2016) Louisiana Rural Definitions. http://www.ers.usda.gov/datafiles/Rural_Definitions/StateLevel_Maps/la.pdf, accessed 20 March 2016.

US Census Bureau (n.d.) *Population demographics for St Tammany Parish, Louisiana in 2016 and 2017*, available at https://suburbanstats.org/population/louisiana/how-many-people-live-in-st-tammany-parish, accessed 27 June 2017.

What Are the 7 Continents (2017) *The 7 continents of the world*, available at http://www.whatarethe7continents.com/north-american-continent/, accessed 11 May 2017.

World Population Review (2017a) Mexico City population 2017, available at http://worldpopulationreview.com/world-cities/mexico-city-population/, accessed 11 May 2017.

World Population Review (2017b) Population of cities in Cuba (2017), available at http://worldpopulationreview.com/countries/cuba-population/cities/, accessed 25 June 2017.

World Weather and Climate Information (2016) Climate Havana, available at https://weather-and-climate.com/average-monthly-min-max-Temperature,Havana,Cuba, accessed 25 June 2017.

WWF (World Wide Fund for Nature) (2012) *Living Planet Report 2012: Summary*, available at http://d2ouvy59p0dg6k.cloudfront.net/downloads/lpr_2012_summary_booklet_final_120505__2_.pdf, accessed 27 June 2017.

6
OCEANIA

Fabricio Chicca and Brenda Vale

Oceania

Brenda Vale

The three stories set in Oceania come from Australia, New Zealand and Tonga. Australia, including small islands lying to the north of the nation of Australia, is the world's smallest continent (Merriam-Webster, 2001: 90), and Oceania is a geopolitical region which includes the islands of the Pacific Ocean. Australia and New Zealand are usually included in Oceania (Dictionary, 2017). To complicate things even more, there is another argument that New Zealand and New Caledonia are the only visible parts of the eighth continent of the world, the largely submerged continent of Zealandia (Campbell and Mortimer, 2014), but this is disputed (Dowding and Ebach, 2017). Here, for the sake of convenience, we refer not to the continent of Australia but to Oceania, which includes Australia and New Zealand, the locations of two of our stories and also includes the Pacific island of Tonga, the setting for the third story here. Given the Pacific Ocean covers approximately a third of the surface of the earth (Facts.net, 2015), there is diversity in both the size and the climates of the places discussed here.

Australia is not the world's largest island because it is defined as a continent. Australia has an area of nearly 7.7 million square kilometres (The World Bank Group, 2017a). With its population of just over 24.5 million (Australian Bureau of Statistics, 2017a), this gives a 2015 population density of 3 people/square kilometre (the same as Iceland and Namibia), compared with 17 for New Zealand and 147 for Tonga (The World Bank Group, 2017b). Of the total Australian population in 2015, 89.4 per cent are considered to live in urban areas (CIA, n.d). This leaves a lot of land virtually unoccupied. As a result, Australia lives within its bio-capacity despite having an average environmental footprint (EF) of 9.3 global hectares/person, of which 4.9 global hectares is the carbon footprint. The bio-capacity is 16.6 global hectares/person (GFN, 2016).

In comparison, New Zealand is a relatively small country that also lives within its bio-capacity. The 2012 EF was 5.6 global hectares/person, and the bio-capacity was 10.1 global hectares/person (GFN, 2016). The total population in 2017 is just under 4.8 million (Statistics New Zealand, 2017). In 2015, 86.3 per cent were defined as urbanised (CIA, n.d.). It used to be said there were 20 sheep for every person in New Zealand, but the declining sheep population means this number is now only six sheep per person, which is still twice that of Australia (three per person) (Statistics New Zealand, 2016). Despite the six sheep and two cows per person (Statistics New Zealand, 2016), in 2009 all the primary industries combined (agriculture, forestry, fishing and mining) formed only 8 per cent of New Zealand's Gross Domestic Product (GDP), of which 4 per cent was from agriculture (Statistics New Zealand, 2012). This compares with the period 2015–2016 in Australia, when agriculture accounted for 2.2 per cent of Australian GDP (Office of the Chief Economist, 2016: 40).

The Kingdom of Tonga has an economy based on agriculture and fishing, both for subsistence and commercially. Tongans living and working overseas are also a major source of income through the money they send home (Bell, 2013). This makes the EF potentially complicated, but accepting that those living overseas have the EF of their adopted country, the EF of Tonga in 2012 was 2.7 global hectares/person (GFN, 2016). This exceeded the bio-capacity of the nation, with a deficit of 1.2 global hectares/person. Besnier (2004) observed that modernisation in Tonga was associated with wealth and this would also apply to EFs of individuals.

Sydney, Australia

Although the largest city, Sydney, is not the capital of Australia (this is Canberra, the site of which was selected for the capital in 1908 [Australian Government, 2015], some 21,000 years after the Canberra area was first home to the Ngunnawal people [Australian Government, 2016]). Sydney is the capital of the state of New South Wales and for many thousands of years, the area was home to the Gadigal people (City of Sydney, 2015). The Europeans arrived in 1788 when Arthur Philip, the city's founder, arrived to set up a penal colony when British convicts could no longer be sent across the Atlantic because of the war between Britain and America (Cities Guide, 2008). From a population of around 1,000 in 1788 (Australia 2016 Population, 2016), the city in 2016 is home to just over 5 million people. This makes it the most populated city in Australia (Australian Bureau of Statistics, 2017b). It also sits on and around the largest harbour in the world, crossed by the famous bridge and home to probably the most recognisable opera house in the world. Completed in 1973, the Sydney Opera House became the most visited place in Australia (Hale and MacDonald, 2005).

Sydney has long been recognised as a sprawling city, forcing those looking for lower-cost housing to move to the outer suburbs (Burnley et al., 1997). More recently, housing affordability has become the issue; in 2009, households with a single moderate income were considered to be unable to buy a median-priced house anywhere in the city (Kupke and Rossini, 2011). For those who can afford to live in Sydney, the climate is sunny, if humid in the summer. The hottest months are January and February, with daily maximum average temperatures of 26°C and daily minimums of 19°C. The coldest months are June and July, where the corresponding average temperatures are 17°C and 8°C (Living in Australia, n.d.).

Near Stratford, New Zealand

The Forgotten World Highway (State Highway 43) leaves Stratford in Taranaki and heads in an easterly direction for Taumarunui, some 148 kilometres distant. A number of small settlements are strung out along this road and our story relates to one of these. The area is rural and remote, as the name of the highway suggests (Gasteiger, 2015).

Stratford is the nearest town. This lies towards the centre of the triangle that forms the region of Taranaki on the east of the North Island of New Zealand. Between Stratford and the sea is Mount Taranaki, an active but currently quiet volcano that was a film double for Mount Fuji in the film *The Last Samurai*. The town's economy is based on agriculture and associated industries, such as meat and dairy processing (Stratford District Council, 2015). In the 2013 census, the population of the whole Stratford area was just under 9,000 people (Statistics New Zealand, n.d.), of which, in 2015, the town itself was home to an estimated 5,600 people (Population.City, 2015).

The climate is temperate but it rains all year, which makes the grass grow and helps explain the popularity of dairy farming. February is the driest month, with just under 100 millimetres of rain, and July the wettest with around 160 millimetres. The warmest month is February, with an average maximum temperature of 20°C and an average minimum of 16°C. The coldest month is July, with respective average temperatures of around 16°C and 4°C (World Weather and Climate Information, 2015).

Nuku'alofa, Tonga

The longest living inhabitant of the capital of the Kingdom of Tonga was Tu'I Malila, the tortoise that was the reputed 1777 present of Captain Cook to the king. This giant tortoise lived for a time in the royal palace until 1966 when, after its demise, it was sent to the Auckland Museum, New Zealand for preservation, before being returned to Nuku'alofa (Robb and Turbott, 1971). The Tongan royal family also earned huge respect when Queen Salote rode through the London rain in an open carriage at the coronation of Queen Elizabeth II (Wood-Ellem, 2001: 244–245).

Nuku'alofa is the largest city in Tonga and is where modern Tonga meets tradition. It is also the first destination for most visitors (Visit Capital City, n.d.).

The capital city, with a population around 25,000, sits on the northern coast of Tongatapu, the largest island of the 36 inhabited islands that make up the country of Tonga (Besnier, 2004).

Nuku'alofa has a tropical climate with plenty of rain all year. February is both the wettest month (225 millimetres) and the warmest month, with an average temperature of 26.3°C, while July is both the driest (90 millimetres) and coolest, with an average temperature of 21.2°C (Climate-Data, n.d.).

Johan and Annie: Surfing in Bondi: Australia

Researcher and writer: Fabricio Chicca

The family

Johan and Annie, both 36 years old, live in Sydney, Australia in a comfortable 135-square metre flat in Bondi close to the beach with their 3-year-old son Billy. Both are trained as civil engineers; however, only Johan works, and not as an engineer, but in the financial department of an international bank. Annie has not worked since Billy was born.

The apartment

Their three-bedroom flat is in a three-storey building made of brick and concrete slabs. The flat was redesigned and renovated in 2012. It has an open-plan kitchen, dining and living room, with timber flooring and marble counters in the kitchen. One of the bedrooms opens to the living room and Annie currently uses it as an exercise space and a playroom for Billy. They feel their flat has all the normal appliances, such as a double-door refrigerator (614 litres), freezer (220 litres), microwaves, blenders, stove, electric oven, electric kettle, juicer, rice maker, bread maker, ice cream maker and toaster. The apartment has four televisions, a 22-inch television in the kitchen, a 50-inch television in the living room, and two 32-inch televisions, one in the parents' bedroom and the other in the playroom.

Transport

The family have two cars, one diesel, one petrol. They plan to sell the diesel car soon as they rarely use it and have to park it on the street as their flat only has one designated parking space. In the last month, they bought 1,300 kilowatt hours of electricity, which they think is a bit higher than normal, and definitely higher than average. This probably happened because the flat gets sun all day in the summer as all windows face north and northwest. This means it gets very hot, and because Billy has a skin condition that gets worse during the summer, they kept the air-conditioning on almost the entire summer. The air conditioner has four outlets, covering almost the entire flat.

Routine

The family routine starts with Johan getting up around 6 am, and either going for a run or surfing. The beach where Johan normally goes in the morning is 250 metres away. During the week of the study, Johan went to the beach on Monday and Tuesday and for a run for the other days as the waves were not good for surfing, which is his priority. After exercising, he comes back to have breakfast with Annie and Billy. He leaves the house at 8 am and walks 1.7 kilometres to the railway station at Bondi Junction, although Annie occasionally gives him a ride if the weather is bad or if he is late. Annie wakes around 7:30 am, or when Billy wakes, and prepares the breakfast. After Johan leaves for work, Annie normally goes for a walk with Billy, either around the local park or along the Sydney Coastal Walk. She normally goes with her friend Norma and her 2-year-old daughter Sonja. In the week of observation, she went for a walk with Norma and her daughter every weekday except Thursday, when she went to Bondi Beach to meet another friend and her mother. All other days, she walked and then had coffee with Norma while they watched Billy and Sonja playing at the playground. She normally returns home around 11 am to feed Billy. After Billy's morning snack, she studies for her online degree in decoration. Around 12:30 pm, she prepares Billy's lunch. On Tuesday and Thursday afternoons, she leaves Billy with her parents, who also live in Bondi Beach, 1.5 kilometres away, and drives to her yoga class, which is another 4.2 kilometres distant. Around 4 pm, she starts preparing the evening dinner. Johan gets back home around 6 pm. He eats lunch out every day and around his workplace there is a great variety of restaurants. Around 10 am at work he always has a spirulina smoothie (750 millilitres) which satisfies him until lunchtime. He enjoys sandwiches with salad and meat followed by another fruit smoothie or cola. Annie normally has a salad and grilled meat for lunch (see Table 6.1).

Social life

As both Johan and Annie grew up in Sydney around the south coast, they have a very active social life with many long-term friends still living in the area. On Monday, after Johan's work, they went to the supermarket for the weekly shop and then stopped by a friend's house for drinks before dinner. They also organised a surfing session for after work the next day.

On Tuesday, as arranged, after work the couple and Billy went to the beach for surfing and hanging out with friends, including Norma and Sonja. They arranged to have a barbecue on Saturday at the public barbecue area at the beach.

On Wednesday, Annie came back from their morning walk and spent the rest of the day looking after Billy and studying for her course. She did not have to prepare a meal, as they were going out for dinner with Annie's sister and her husband, also in Bondi. Johan arrived home at 6:15 pm and went to the beach for an unsuccessful surfing session, after which, around 7:30 pm, they drove to Bondi to have dinner with Annie's sister. They returned home at 11 pm, much later than they were

TABLE 6.1 Food

		Billy	Annie	Johan
Monday	Breakfast	Ella's Organic Food—Strawberry and cream	Bowl of fruit salad (pineapple, papaya, orange, apple and banana), yoghurt (120 grams) and an espresso	Four slices of toast with beans (120 grams), two scrambled eggs with bacon (200 grams), 400 millilitres of orange juice and an espresso
	Snack	Banana and a piece of bread	Mocha	Spirulina smoothie (750 millilitres)
	Lunch	Grilled chicken and grilled potatoes	Grilled chicken breast and tomato salad	Tuna melt sandwich, an apple and water
	Snack	Piece of bread and cheese	–	750 millilitres BOC juice (beetroot, orange and carrot)
	Supper	A steak (400gms each for the parents and a small piece for Billy) and mashed potatoes and peas		
Tuesday	Breakfast	Papaya, scrambled egg and chocolate milk (200 millilitres)	Half a papaya, scrambled egg and bacon	Three slices of toast with vegemite, two scrambled eggs with bacon (200 grams) and 400 millilitres of orange juice
	Snack	Piece of bread and chocolate milk (200 millilitres)	Mocha and a croissant	Spirulina smoothie and an espresso
		Snack		
	Lunch	Half a mince pie and grape juice (100 millilitres)	Greek salad (small container) and a mince pie	Pastrami sandwich, coleslaw and a Coke (330 millilitres)
	Snack	Half a mince pie	–	
	Supper	Bolognese pasta and chocolate ice cream		Spirulina smoothie and a nut bar

Wednesday	Breakfast	Toast with Vegemite and chocolate milk	Two slices of toast with Vegemite, a glass of milk (400 millilitres) and an espresso	A glass of milk (late for work)
	Snack	Croissant	Mocha and a croissant	Spirulina smoothie and two nut bars
	Lunch	Rice and chicken	Croissant and chocolate milk (400 millilitres)	Seafood fried rice
	Snack	Piece of cheese and three crackers	—	—
	Supper	Eat out—large mushroom pizza and a small margherita, with a bottle of red wine, panna cotta and tiramisu		
Thursday	Breakfast	Banana and toast with Vegemite and chocolate milk	Two slices of toast with Vegemite, a glass of milk (400 millilitres), half a papaya and an espresso	Skipped (late for work)
	Snack	Guava juice	Mince pie and guava juice	Spirulina smoothie
	Lunch	Smashed potatoes and chicken	Packet mushroom soup (600 millilitres) and two slices of toast	Cowboy sandwich (roast beef with cheese and tomatoes) and a Coke (300 millilitres)
	Snack		Snack chips (30 grams)	An espresso and a Danish
	Supper	Roast chicken, steamed broccoli and smashed potatoes		
Friday	Breakfast	Toast with Vegemite and chocolate milk	Two slices of toast with Vegemite and an espresso	Four slices of toast with Vegemite, chocolate milk (400 millilitres) and an apple
	Snack	Croissant and a box of tropical juice	Mocha and a cheese baguette	—

(Continued)

TABLE 6.1 (*Continued*)

		Billy	Annie	Johan
	Lunch	Broccoli and chicken	—	Pastrami sandwich, medium size quinoa salad and berry juice
	Snack	Apple	—	Mince pie and orange juice (300 grams)
	Supper	Eight lamb chops with potatoes and gravy		
Saturday	Breakfast (eaten out)	French toast (half portion)	Eggs Benedict and a mocha	Smoked salmon and toast, orange juice and a long black
	Snack	Apple and a box of tropical juice box	—	—
	Lunch	Half a mince pie	Mince pie	Ricotta and spinach pie
	Snack	Banana, chocolate milk and a cookie	—	—
	Supper	BBQ at beach		
Sunday	Breakfast (eaten out)	Toast with jam and chocolate milk	Mocha and fruit salad with granola and yoghurt	Big Breakfast (bacon, two eggs, grilled tomato, potato rösti and two sausages)
	Snack	Banana and milk	—	—
	Lunch (eaten out)	Piece of vegetarian pizza, chocolate ice cream	Vegetarian pizza	Four-cheese pizza, Diet Coke and chocolate mousse
	Snack	Apple	—	—
	Supper	Roasted vegetables and roast chicken (630 grams between them)		

expecting. As a result, next morning, Johan was late up but even then went for his run, skipping breakfast.

On Thursday, Annie's mother looked after Billy for the afternoon, so Annie managed to go to yoga and finish her study assignments. She picked Billy up at 4:30 pm and went straight home to prepare the dinner. After dinner, they had a lazy night watching a movie on the television. On Friday, Johan went for a run before work. Annie had her walk with Norma and Sonja in the morning, came home and fed Billy and then worked on her assignment whenever Billy allowed, before preparing the dinner. After dinner, the family walked to Annie's mother's house in Bondi. They got back home around 10 pm, watched a movie and went to bed.

Weekends are normally a lazy time for the family. Johan, who wakes up early, looks after Billy until Annie wakes up. That Saturday, she got up at 8:30 am, freeing Johan to go the beach for a surfing session. As the waves were not suitable and the beach was overcrowded, he headed home and they went out for brunch with Norma and her family. They got back home around 2 pm, just in time for Billy's nap. While Billy was in bed, Johan went to the supermarket to buy meat (2 kilograms of lamb chops and six steaks) for the barbecue, 12 bottles of beer, a bottle of white wine and a cake. They went out for the barbecue at 5 pm, which was a bit too late to get a good barbecue spot at the beach. The group then drove south to Coogee Beach, where they got a prime spot by the sea. They returned home around 10 pm, watched a movie and went to bed. On Sunday, as on Saturday, Johan waited for Annie to get up and, instead of going for a run, drove to Bondi Beach to surf. Annie then drove down to Bondi to have brunch with Johan (so that day, they used both cars). They met at 11:30 am and had their brunch with Annie's parents. After this, they returned home so Billy could take his nap. When Billy woke, he and his father went out to a friend's house (250 metres away) to watch a rugby game so Annie could clean the house and finish her assignment. After the game, they went to a rotisserie for a take-away dinner. The family ate this together at 8 pm and after that watched a movie before bed. During the week, they drove 22 kilometres with the petrol car and 13 kilometres with the diesel.

Michael and Georgie: Cows eat grass all year round: New Zealand

Researcher and writer: Fabricio Chicca

The family

The family live in the New Zealand countryside, in a farm 17 kilometres east of Stratford in Taranaki, in the North Island. They own and operate a dairy farm of 54 hectares. Dairy farming is very common in the region. Their farm is a roughly flat area, and is perfect for cattle. The family consist of husband (Michael) and wife (Georgie) (41 and 37 years) and their two girls of 12 and 10.

The house and heating

The family live in a 130-square metre typical New Zealand house, timber-framed and timber clad, with three bedrooms, two bathrooms and an open-plan kitchen, living and dining space. Very recently, an extension has been completed. This consists of a conservatory to increase the living space and a small office/study. The house is fully insulated, and except for windows in the bathrooms and over the kitchen sink, has double-glazing throughout. The roof finish is corrugated steel. The open-plan living space has a wood-burner and a heat pump. The children's bedrooms are heated by electric panel heaters, while the parents' room has an electric oil-filled convection heater. They consider their house very warm but very small. During the week described, the family used around 0.5 cubic metres of pine logs for the wood-burner. They are planning a new extension, which will include a guest room and the conversion of the second bathroom into an en-suite bathroom for the parents.

Vehicles and appliances

The family own two cars. One, a station wagon, is considered to be the family car, while the other, a small Japanese "ute" (the NZ name for a pick-up) is a work vehicle. A number of additional vehicles are used for work on the farm, including a tractor and two quad bikes.

The household electricity use changes according to the time of the year, as the demand for heating is more intense during the winter. In October 2016, the time of this description, the family used 911 kilowatt hours, which is much higher than the same time a year ago when they only used 601 kilowatt hours. They feel this year is much colder than the same time a year ago. The family have all modern amenities, with two televisions (40 inches and 26 inches), a refrigerator and all the basic kitchen appliances, one laptop and two tablets, a stereo and a satellite antenna for internet and television. Like many places in rural New Zealand, the house is not served by reticulated water and sewerage. Water comes from a certified well and sewage disposal is via a septic tank.

Routine

The dairy tasks produce a very tough family routine. In addition, the current dairy market situation is affecting the family finances. The milk price has fallen around 25 per cent in the past 4 years (Dairy NZ, 2015:49). A dairy farm requires a great deal of work, and because of the fall in the price of milk, the farm now has no employees and all the work has to be done by the family themselves. The daily routine starts at 4 am, when Michael gets up to gather the cattle and bring them to the milking shed for the first milking of the day, which can all take up to 3 hours. By the time he returns to the house, the rest of the family are organising themselves for the day. The girls are dressing and Georgie is setting breakfast and making their packed lunches.

After breakfast, the girls head off to take the bus to their school in Stratford, which is the nearest town, and Michael helps with the kitchen clean-up. After that, he talks briefly to Georgie about the rest of the workday. During the week described, the couple worked on two main tasks. The first was the replacement of the southern part of the fence, which had been partially destroyed some weeks before during a storm. The second, less urgent, task was finishing the painting of the milking shed and tool shed. The milking shed's roof needs repair, but they will do that in the summer. The post-breakfast meeting is to decide what they will do that day and how they will do it. In the week in question, all days except one were dedicated to the fence replacement, as the weather was rainy. Painting only happened on the one fine day in the week. After the decisions are made, the work starts. Both work on the fence, but after an hour, Georgie comes home to prepare the lunch, returning around 11 am to pick up Michael. They would normally each use a quad bike, but one of the bikes was not working at the start of the week. During the week, the working quad bike travelled 56 kilometres and the second one 4 kilometres on Sunday, after it was fixed. As the weather was rainy during the week, they came back home every day for lunch. After lunch, however, the tasks were different. On Monday, Michael was feeling unwell, and went to bed until the second milking of the day at 4 pm. On Monday and Tuesday afternoons, Georgie prepared meals to be frozen and eaten during the week. On Tuesday afternoon, Michael had a meeting in the nearest settlement with another milk producer.

On Wednesday, they both worked on the damaged fence until 4 pm, when they returned for Michael to do the second milking and Georgie to make dinner. On Thursday, they worked until lunch, which they had at home, and then went to New Plymouth, 40 kilometres away. Georgie had a doctor's appointment and Michael was after spare parts so he could mend the second quad bike. They also went to the supermarket for a big shop. These visits to New Plymouth, which is the largest city in the region, with 74,184 inhabitants (Statistics New Zealand, 2016), happen twice a month, mostly for shopping. They also bought supplies for the farm, arriving back home around 3 pm in time for the second milking of the day. Wednesdays and Fridays are the days when the children have after-school activities (swimming and dancing) and must be picked up in Stratford. That day, Georgie drove to town to pick up the girls and Michael prepared the dinner.

The family normally have dinner together. This is a ritual that is very much appreciated by the parents, who repeatedly referred to it as the best part of the day. The girls enjoy it too. They often take part in the dinner preparation and occasionally take over the entire process, leaving the parents to do the clearing up afterwards. They try to have dinner every day around 6 pm so Michael can catch the sports session on the nightly television news. After this, the girls take control of the television and choice of programmes. Since 2014, the amount of internet usage at home has been controlled, and as no-one has a smartphone and the house has only a work computer and two tablets, the parents feel they have been successful. The children are allowed to use the tablets for homework and for an hour a day for their own purposes, and this they normally do at night.

Thursdays and Saturdays are special days in the routine. Because the cows have to be milked twice every day, the family used to have no breaks, but this changed a couple of years ago, when Michael and his father, who also has a dairy farm, made an arrangement for sharing their respective milking sessions. On Thursdays, Michael has a double milking day. He starts an hour earlier at his farm, milks the cows, and instead of coming back home for breakfast, drives to his father's farm and milks his cows. The same routine happens in the afternoon, so that day the family have a slightly later dinner. On Saturdays, Michael's father does all milking on both farms, freeing up a day for the family. On Saturday, therefore, Michael gets up later than usual, around 8 am, and sets the breakfast table, although everyone is responsible for their own breakfasts and clearing up afterwards. One by one the family get up and have breakfast. During the week described, the family had a very busy weekend. On Saturday, Georgie and the girls went to Stratford to buy a gift for a neighbour who was throwing a barbecue anniversary party on Sunday and to buy food for a special dinner for Michael's parents. While out, they had lunch of meat pie and ice cream. While the women were in town, Michael drove to a friend's farm for a chat and a beer, coming back around 3 pm to do the second milking of the day instead of his father doing this, by which time Georgie and the girls were also home. The whole family then helped. The younger daughter is keen to help her father with the farm tasks, and whenever she is home, she volunteers to help. On Sunday, after the first milking of the day, they all went to the barbecue, returning just in time for the second milking.

Food

The family felt they have an average diet for New Zealand, although were unsure what exactly that meant. In the morning, Michael eats what he calls his first breakfast, which consists of a boiled egg (boiled the night before), a glass of milk and two slices of toast with butter. On the Thursday, he also had a banana and a bowl of porridge. His goal is to join the family for their first and his second breakfast. However, lately it has been harder to achieve the common breakfast and he is losing this habit. For breakfast, the girls have cereal with milk or yoghurt, a piece of cheese and a glass of orange juice. The older daughter also has a banana and toast with strawberry jam. During the week, Michael only managed to join the family twice, when his second breakfast was beans, toast and avocado, with a mug of coffee with milk. Georgie's breakfast is normally buttered toast and coffee. While the children are eating, she prepares their school lunches, which always consist of a sandwich with cheese, some kind of meat (salami, pepperoni, turkey or chicken), a small bottle of juice and a piece of fruit.

The lunches for Michael and Georgie had more variety than breakfast. On Monday, they each had a couple of salami and cheese sandwiches with an apple and milk. Tuesday and Thursday they had a steak with mashed potatoes, and on Wednesday, in town, they both had a combo from an international chicken fast-food chain. On Friday, they had pasta bolognese. On Sunday, as noted above, the family went to a friend's house in Opunake for a barbecue. The family dinner, which they consider the most important meal, is even more diverse. On Monday night, they had red meat with rice and fried tomatoes and homemade bread, and on Tuesday, steaks, mashed potatoes and asparagus. Wednesday night, Michael had a steak with cheese pasta, the girls had lamb chops and Georgie had a tuna salad. Thursday night they all had red meat and bacon with fried vegetables, and on Friday, beef curry and rice. On Saturday, the family and Michael's parents had ham with gravy. On Sunday, after the barbecue lunch, they ate the leftovers from Saturday's dinner.

Social life

The family state their lack of social life is a consequence of the daily routine that comes from working in the dairy industry. The drop in the milk price has pushed them to the limit so they can no longer afford any paid help, such as an occasional milk relief service. The farm is not big enough to have an employee and not small enough to allow free time. When they had more spare money, when milk prices were better, milk relief was often hired and the family used to have more social contact, especially with their close neighbours. Lately, even weekend rugby and barbecues with neighbours have been put aside as they have no time or energy to join in.

Travel and food

During the week, the family drove 60 kilometres in total using the quads, 145 kilometres in the ute and 111 kilometres in the family car. During this time, the family diet included 9.5 kilograms of meat, 2.5 kilograms of cheddar cheese, 12 litres of milk, 4 litres of orange juice, 400 grams of rice and 4.5 kilograms of potatoes.

Tane and Kamala: Saving hard: Tonga

Researcher and writer: Fabricio Chicca

The extended family

Tane and Kamala and their two boys, aged 3 and 5 years, live on the outskirts of Nuku'alofa, the capital of Tonga, with members of their extended family. During the time the story was written, the extended household included two cousins (19 and 36 years old) and Kamala's auntie. Tane is 31 years old and works in an office in Nuku'alofa. Kamala is 27 and works 3 days a week at the national airport.

The house

The family live in an 86-square metre timber-frame house, with a corrugated metal roof. The house is elevated above ground by 1.2 m. The principal space is the living and kitchen area where the family spends most of their time. Another room is for the couple and another would normally be used by the children for sleeping, but this room has been given to Kamala's auntie. Both rooms open off the living room. Both cousins and the boys currently sleep in the living room. The house was bought by the husband's parents when they came back from New Zealand in 1990. They had lived there for 11 years, although they never got used to the New Zealand weather.

Appliances, energy and vehicles

The house has the basic appliances of a small refrigerator (180 litres), a 26-inch television with a DVD player, stove, blender, three pedestal fans and a stereo. Three years ago, with the help of a non-government organisation (NGO) and the local government, the family acquired a solar water heater from Thailand that provides up to 120 litres a day of hot water on sunny days. They also have three water tanks capable of storing 2,400 litres of rainwater. The family use around 250 kilowatt hours of energy a month. The family replace their 13-kilogram gas bottle every 45 days. They have a 20-year-old Japanese car, but the car had not been going for the past 2 months as the family were waiting for a spare part to be delivered from New Zealand. Without the car, they were relying on buses to get around.

The house site

The house sits on a 1,120-square metre site which has been divided into two. The main area, where the family's house is located, has direct access to the street. Tane's parents live in a small, very old house from the 1940s at the rear of the site, separated by a precarious fence but with direct access to the main house. The family are building another construction in the middle of the site to host family events and parties, as currently everyone gathers in the main house. The family also raise chickens (around 50 at present). Despite the fact that the family are considered "high middle class", they have a very modest lifestyle. For the past 5 years, they have been saving money to buy a better house and to visit parents in New Zealand, and this has directly impacted their lifestyle.

Routine

During the working week, the family have a very tight schedule, particularly when Kamala goes to work (Mondays, Wednesdays and Thursdays). In those days, Tane and Kamala have a 20-minute walk to the crossroads to catch the bus. There, they go in different directions. Tane heads north towards the western bus terminal (5 kilometres), from which it is another 5-minute walk to his workplace. Kamala gets a tourist bus to the international airport 18 kilometres away, although occasionally she gets a lift to work with some of her workmates. On the week described, she had a lift to and from work on Thursday. Car sharing was part of the family routine, but as their car is not working, they have been unable to do their part in this. Tane also got two lifts to work in the week in question.

Both could have had more but felt embarrassed to accept because they could not reciprocate.

While Tane and Kamala are at work, the boys are looked after by their grandparents or auntie, who is not currently working. The day starts at 7 am when Tane and Kamala get up. Kamala prepares the breakfast if there are no leftovers from the night before. The family normally have fruit, watermelon or mangoes, and a fried pancake called *Keke Vai*. During the week auntie prepared *Keke Vai* once, and another day the family shared corned beef for breakfast. On all other days, they ate leftovers from the previous dinner, except Sunday when they fasted. The family always have their big meal at night. After this, the men sit outside for a smoke and a chat while the women do the cleaning. When this is done, the women either join the men or they all sit to watch television or a movie. They go to bed around 10 pm every night, and are not in the habit of going out.

On Monday, the family had leftovers from the Sunday dinner. Tane and Kamala took food to work for lunch, and those at home ate canned meat and boiled taro with homemade coconut milk and fried onions with fried eggs. That night, Tane brought in a large fish and some sea snails and the entire extended family ate together. The fish was prepared by the women using the traditional underground oven, the *umu*, and it was served with boiled taro and coconut, with watermelon for dessert. On Tuesday, they ate taro with coconut milk and boiled eggs in the morning. Tane took taro and a can of tuna for his lunch. At home, Kamala and auntie prepared a large meal using cassava (which is grown by the grandparents in a small plot 250 metres away from their house) with coconut milk, onions, eggs and a chicken stew (three chickens were slaughtered). The meal was served for lunch, dinner and breakfast and a small portion was used for the parents' lunches. On Wednesday, the boys, auntie and cousins went to the grandparents' house where they had sweet potato with corned beef and eggs. For dinner, Tane brought home mutton flaps which the family ate with cassava. The mutton was also prepared in the *umu* and the grandparents joined them for the meal. On that occasion, there were no leftovers for breakfast, which was when Tane and Kamala had the corned beef. A small portion of cassava was boiled and, with some boiled eggs, this made their lunches to take to work. That evening, Kamala met Tane from work and they went together to the traditional Tongan market. There they bought fish and taro leaves. At home, auntie had bought mussels and the grandparents, seaweed. The meal was prepared in a relative's house down the street using their larger *umu*. This meal was shared by a large number of relatives who also contributed food to the meal. Gatherings like this are relatively normal in Tonga.

On Friday, the family had sweet potato and eggs for breakfast. Tane took some for his lunch, with the addition of some coconut and a piece of chicken left over from the night before. Kamala and the boys went to town (5 kilometres) for a medical appointment. They skipped lunch, but ate bananas she bought in the market. They later met Tane and went home together. That night, grandmother prepared the meal of cassava and eggs with barbecued chicken. It was a very small gathering as only the family and grandparents shared the meal.

The weekend

Weekends are almost entirely dedicated to church. On Saturday, they got up at 7 am and had cassava with coconut milk, bananas and watermelon, and boiled some eggs to bring to the church. The entire family went to church, which is about 3 kilometres away, to join in events and help with the church activities. They either walk for 45 minutes or take a lift with relatives. During the week described, the entire family, including grandparents, auntie and cousins, were picked up by a relative. Kamala is a volunteer in the church food charity group while Tane organises events for the youth and also conducts the church's band and choir. Grandmother weaves at the church while grandfather hangs out with his friends. That Saturday, the family were invited for a Tongan barbecue to celebrate the anniversary of a relative who was turning 80. They had their tasks at the church shortened and headed to the relative's house at 2 pm (1 kilometre away). They spent the rest of the afternoon there. It was a traditional Tongan feast, with fish, chicken and pork cooked either in the traditional *umu* or barbecued in the Western style. The family came back home accompanied by a number of relatives and watched television before going to bed.

Although it is not very common in Tonga, the family practices a Sunday fast. They only eat after the church service. They wake up at 7 am and prepare themselves for church. They wear their best clothes. On a normal Saturday, after the church tasks, they go to the market and buy food for Sunday. However, due to the anniversary feast, they had a slightly different routine on Sunday, as they had nothing to eat and the shops in Tonga are all closed on Sunday. After church, Tane went to a relative for some fish for the Sunday meal, while the rest of the family headed home. When he got home, the *umu* was already prepared and the fish went in almost immediately. Kamala had prepared sweet potato with coconut milk and spices, and grandmother had prepared taro leaves, and corned beef wrapped in banana leaves to go into the oven with the fish. The family had their meal at 4 pm, and the taro leaves were spared for a small meal at night, breakfast and Tane and Kamala's Monday lunches.

References

Australia 2016 Population (2016) *Population of Sydney in 2016*, available at http://australia population2016.com/population-of-sydney-in-2016.html, accessed 29 June 2017.

Australian Bureau of Statistics (2017a) *Population clock*, available at http://www.abs.gov.au/ausstats/abs%40.nsf/94713ad445ff1425ca25682000192af2/1647509ef7e25faaca2568a900154b63?OpenDocument, accessed 28 June 2017.

Australian Bureau of Statistics (2017b) *Regional population growth, Australia, 2015–2016*, available at http://www.abs.gov.au/ausstats/abs@.nsf/mf/3218.0, accessed 29 June 2017.

Australian Government (2015) History of the capital, available at https://www.nationalcapital.gov.au/index.php/education/history-of-the-capital, accessed 29 June 2017.

Australian Government (2016) Canberra—Australia's capital city, available at http://www.australia.gov.au/about-australia/australian-story/canberra-australias-capital-city, accessed 29 June 2017.

Bell A V (2013) The dynamics of culture lost and conserved: Demic migration as a force in new diaspora communities, *Evolution and Human Behavior*, 34(1), pp. 23–28.
Besnier N (2004) Consumption and cosmopolitanism: Practicing modernity at the second-hand marketplace in Nuku'alofa, Tonga, *Anthropological Quarterly*, 77(1), pp. 7–45.
Burnley I H, Murphy P A and Jenner A (1997) Selecting suburbia: Residential relocation to outer Sydney, *Urban Studies*, 34(7), pp. 1109–1127.
Campbell H and Mortimer N (2014) *Zealandia: Our Continent Revealed*, Auckland: Penguin Books.
Central Intelligence Agency (CIA) (n.d.) *The World Factbook*, available at https://www.cia.gov/library/publications/the-world-factbook/fields/2212.html, accessed 29 June 2017.
Cities Guide (2008) *Sydney history*, available at http://go.galegroup.com.helicon.vuw.ac.nz/ps/i.do?&id=GALE|A177329394&v=2.1&u=vuw&it=r&p=ITOF&sw=w&authCount=1#, accessed 29 June 2017.
City of Sydney (2015) Aboriginal history, available at http://www.cityofsydney.nsw.gov.au/learn/sydneys-history/aboriginal-history, accessed 29 June 2017.
Climate-Data (n.d.) *Climate: Nuku'alofa*, available at https://en.climate-data.org/location/54130/, accessed 30 June 2017.
Dairy NZ (2015) New Zealand Dairy Statistics 2015–16, available at https://www.dairynz.co.nz/media/5416078/nz-dairy-statistics-2015-16.pdf, accessed 16 December 2017.
Dictionary (2017) Oceania, available at http://www.dictionary.com/browse/oceania, accessed 28 June 2017.
Dowding E M and Ebach M C (2017) Geography: Zealandia is not a continent, *Nature*, 543(7644), p.179.
Facts.net (2015) *Pacific ocean facts*, available at http://facts.net/pacific-ocean/, accessed 28 June 2017.
Gasteiger A (2015) The forgotten world highway, *New Zealand Geographic*, available at https://www.nzgeo.com/stories/the-forgotten-world-highway/, accessed 30 June 2017.
Global Footprint Network (GFN) (2016) *National Footprint Accounts 2016 Edition*, available at https://data.world/footprint/nfa-2017-edition, accessed 28 June 2017.
Hale P and MacDonald S (2005) The Sydney Opera House: An evolving icon, *Journal of Architectural Conservation*, 11(2), pp. 7–22.
Kupke V and Rossini P (2011) Housing affordability in Australia for first home buyers on moderate incomes, *Property Management*, 29(4), pp. 357–370.
Living in Australia (n.d.) *Sydney's climate*, available at https://www.livingin-australia.com/climate-weather-sydney/, accessed 29 June 2017.
Merriam-Webster (2001) Merriam-Webster's Geographical Dictionary, Springfield, MA: Merriam-Webster, Inc.
Office of the Chief Economist (2016) *Australian industry report 2016*, available at https://www.industry.gov.au/Office-of-the-Chief-Economist/Publications/AustralianIndustryReport/assets/Australian-Industry-Report-2016-Chapter-2.pdf, accessed 29 June 2017.
Population.City (2015) Stratford: Population, available at http://population.city/new-zealand/stratford/, accessed 30 June 2017.
Robb J and Turbott E G (1971) Tu'i Malila, "Cook's Tortoise", *Records of the Auckland Institute and Museum* 8, pp.229–233.
Statistics New Zealand (n.d.) *2013 Census QuickStats about a place: Stratford District*, available at http://www.stats.govt.nz/Census/2013-census/profile-and-summary-reports/quickstats-about-a-place.aspx?request_value=14186&tabname=, accessed 30 June 2017.
Statistics New Zealand (2012) What New Zealand actually does for a living: From manufacturing to a services-oriented economy, available at http://www.stats.govt.nz/

browse_for_stats/economic_indicators/NationalAccounts/Contribution-to-gdp.aspx, accessed 29 June 2017.

Statistics New Zealand (2016) New Zealand is home to 3 million people and 60 million sheep, available at http://www.stats.govt.nz/browse_for_stats/population/mythbusters/3million-people-60million-sheep.aspx, accessed 29 June 2017.

Statistics New Zealand (2017) Population clock, available at http://www.stats.govt.nz/tools_and_services/population_clock.aspx, accessed 29 June 2017.

Stratford District Council (2015) Stratford Economic Development Plan 2012–2015, available at http://www.stratford.govt.nz/images/Reports/SDC%20Economic%20Development%20Strategy%20-%20August%202012.pdf, accessed 30 June 2017.

Visit Capital City (n.d.) *Nuku'alofa, capital city of Tonga*, available at http://www.visitcapitalcity.com/oceania/nuku-alofa-tonga, accessed 28 June 2017.

Wood-Ellem E (2001) *Queen Sālote of Tonga: the story of an era, 1900–1965*, American Council of Learned Societies, Humanities E-Book: URL http://www.humanitiesebook.org/.

The World Bank Group (2017a) *Land area (sq km)*, available at http://databank.worldbank.org/data/reports.aspx?source=2&series=AG.LND.TOTL.K2&country=, accessed 28 June 2017.

The World Bank Group (2017b) Population density (people per sq. km of land area), available at http://data.worldbank.org/indicator/EN.POP.DNST, accessed 28 June 2017.

World Weather and Climate Information (2015) Average monthly weather in Stratford, New Zealand, available at https://weather-and-climate.com/average-monthly-Rainfall-Temperature-Sunshine,stratford-taranaki-nz,New-Zealand, accessed 30 June 2017.

7
SOUTH AMERICA

Emilio Garcia, Fabricio Chicca and Brenda Vale

Introduction

Brenda Vale

Most of the continent of South America lies within the southern hemisphere, with the exception of Venezuela, Guyana, Surinam and French Guiana that, with most of Colombia, are all in the northern hemisphere. The continent is fourth largest in terms of area (after Asia, Africa and North America) and fifth in terms of population (after Asia, Africa, Europe and North America) (What Are the 7 Continents, 2017). Of the 12 independent countries that call South America home, the official language of one is Portuguese (Brazil), of nine Spanish (Argentina, Bolivia, Chile, Colombia, Ecuador, Paraguay, Peru, Uruguay, Venezuela), one Dutch (Surinam) and one English (Guyana). The three dependent territories are French Guiana, South Georgia and the South Sandwich Islands (which count as one) (United Kingdom) and the Falkland Islands (United Kingdom, but Argentina are not happy about this). These languages reflect a long history of European conquest and colonisation.

Brazil is both the largest nation in South America and the most populous, occupying 50 per cent of the area of the continent, and it is home to 52 per cent of South Americans (Countries of the World, 2017). The main continental land mass stretches from just above the equator south to Cape Horn, with a number of islands lying between the Cape and the Antarctic Circle. The climates thus vary from very hot to very cold. Cold weather is also found in the mountain plateau of the Andes, which is the longest continental mountain range, running for 7,242 kilometres along the western edge of the continent from Venezuela in the north to Argentina in the south (Zimmermann, 2013). South America is also home to the Amazon, now claimed to be the world's longest river (6,840 kilometres compared with the 6,695 kilometres of the River Nile [United Press International NewsTrack, 2007]), and the largest area of rainforest, which is still under threat from human activity (Schiffman, 2015).

The average environmental footprints (EFs) of South American nations reflect the varying lifestyles within them, as the stories show. However, most South American EFs are much closer to one-planet (i.e. fair earth share) living than those of North America or much of Europe. Brazil, with its large population, has a 2012 average EF of 3.1 global hectares/person, the same as Argentina. Chile is higher, at 4.4 global hectares/person, and Colombia lower at 1.9 global hectares/person, with other countries lying in between (GFN, 2016). There is no particular relationship between national average EF and the percentage of people living below the poverty line (Table 7.1). All countries live within their bio-capacity, with the EF of Ecuador just equal to its bio-capacity (Table 7.1).

Tucumán, Argentina (written by Emilio Garcia)

Tucumán is one of the 23 provinces that constitute the Republic of Argentina. Located in the northwest of Argentina, with the foothills of the Andes to its west and a flat fertile plain to the east, Tucumán is the second smallest province in the country (Encyclopaedia Britannica, n.d.). Since its foundation in 1565, the province has been the political, economic and cultural centre of the northwest region, with sugar cane, introduced in the nineteenth century, being the main industry (Encyclopaedia Britannica, n.d.). The provincial capital in the centre of the province is the city of San Miguel de Tucumán, confusingly also known simply as Tucumán. Its growing population in 2010 was just over 1.5 million people (Knoema, 2017),

TABLE 7.1 EFs of South American nations

Country	2012 EF in global hectares/person*	2013 Bio-capacity in global hectares/person**	Percentage population below the poverty line***	Year of estimate
Argentina	3.1	6.8	30	2014
Bolivia	3.0	16.9	45	2014
Brazil	3.1	8.8	21	2014
Chile	4.4	3.7	15	2009
Colombia	1.9	3.7	33	2012
Ecuador	2.2	2.2	26	2013
French Guiana	2.3	106.2	N/A	N/A
Guyana	3.1	69.5	35	2006
Paraguay	4.2	11.5	35	2010
Peru	2.3	3.9	26	2012
Surinam	4.3	89.4	70	2002
Uruguay	2.9	10.2	19	2010
Venezuela	3.6	2.7	32	2011

*(GFN, 2016)
**(GFN, 2017)
***(Index mundi, n.d.)

with a population density of 64.3 persons/square kilometres, making it the fifth most densely populated city in Argentina (Population Labs, 2011).

Tucumán has a subtropical climate with dry winters and rainy summers. Winter lasts 2 months with temperatures that can be close to 0°C. Autumn and spring are short and temperate but summer feels like an eternity, with high humidity and temperatures above 40°C, even at midnight. In the days before air conditioning became common, people used to work from 8 am to 1 pm, stop to have lunch and a siesta and wake up around 4 pm, and then come back to work until 8:30 pm. However, in recent years, there has been an increasing change in the working hours and more people are working from 8 to 5 pm without a siesta. Because of these working traditions, Tucumán has always had a very busy nightlife. Young people and students tend to go out for dinner or drinks from Thursday to Sunday and come back home at 3 am. Families go out during the weekends, when playgrounds are open until 2 am.

For centuries, life in Tucumán has revolved around the traditional core of the city, a typical Spanish grid of nine by nine blocks with a main square in the middle and four avenues around the edges. Everybody used to live and work within the city boundaries, with rich people living closer to the central square and poor people occupying the periphery. After the 1960s, middle- and high-class families started to move outside the traditional centre and gated communities evolved (Malizia, 2015).

Dourados, Brazil

Dourados lies far to the west and slightly north of Rio de Janeiro, close to the border with Paraguay. The city of Dourados has its origins in the 1861 Military Colony of Dourados which was established at the time of the invasion by Paraguay. Before then, the area was home to the Terena and Kaiwa tribes, and the indigenous population is still significant (Dourados Prefeitura, 2017a; Tavares, 2015). In the 2012 census, the city itself had a population of just over 200,000. The area is agricultural and known for the production of soya beans and corn as well as cattle ranching (Dourados Prefeitura, 2017b). Our story comes from a family farming on the outskirts of the city.

The climate of Dourados is tropical, with most rainfall in the summer, especially in March and April. The annual average rainfall is 1,100 millimetres. The average annual temperature in Dourados is 26.0°C, with little variation between the hottest month of December, with an average 26.8°C, and the coldest of July, at 25.0°C (Climate-Data, n.d.).

São Paulo, Brazil

São Paulo, the capital of the state of the same name, lies to the west and slightly south of Rio de Janeiro and, with a population of 20 million, is the largest city in Brazil (Perrotta-Bosch, 2016). Strictly speaking, the population refers to the São Paulo Metropolitan Region (SPMR). Because of its size and population, SPMR is known for its traffic jams. In the 2009 census, around 55 per cent of commuters had a home-to-work journey of more than 30 minutes, and 23 per cent (some 1.8 million people) more than 60 minutes (Neto et al., 2014). By 2014, another survey found the average total time per day spent in traffic, whether stuck or moving, was 2 hours 46 minutes (NOSSASP, 2014).

São Paulo was established as the Bandeirantes, a type of militia, pushed inwards from where the Jesuits had settled on the coast in the fourteenth century, after Afonso de Souza had claimed the area for Portugal. The Bandeirantes were in search of gold and precious stones (Brazilian Travel Information, 2015). By the 1980s, São Paulo was the fourth largest city in the world. Since then, not only does the core contain more people but the outer suburbs have spread, with much of the suburban growth happening in the informal settlements of *favelas*, which house 20–30 per cent of the population. *Favelas* also grow up on flood plains and hillsides, or in other words, on land that is not deemed suitable for conventional buildings (Earth Observatory, 2014).

The Tropic of Capricorn passes through São Paulo and this location, combined with its elevation above sea level, gives it a temperate climate. July is the coldest month, with an average temperature of 14°C, and February the warmest, with an average of 21°C. There is plenty of rain, at 1,422 millimetres annually, with most falling over the summer (Encyclopaedia Britannica, 2017).

Valentina and Joaquin: Family interaction in Tucumán

Writer and researcher: Emilio Garcia

The family

The family described in this chapter belong to the new generation that lives in gated communities but whose parents and work are in the city centre. The family are Valentina and Joaquin; their baby; Joaquin's son, Marcos, from his first marriage (Marcos lives in the house a couple of days each week); their maid, Lucia, who sleeps in the house during the week; and the family dog. The family belong to the upper middle class of Tucumán. The parents are both lawyers, working in public administration. Together they have a monthly salary of 70,000 pesos (US$7,000), which represents almost nine times the minimum wage in Tucumán of around 8,500 pesos (US$850) per month. Within the family group, it is important to include the grandparents with their maid, because without them the routine of the family would be incomplete.

The house

The family lives in a gated community 12 kilometres from the city centre. The house is 298 square metres and was designed by an architect and built 3 years ago. It has four bedrooms and four bathrooms, and it is on two levels. The plan is a rectangle with a longer east–west axis to allow the main rooms to be open to the sun. In Tucumán, houses, including this one, are made from brick with a seismic-resistant structure of concrete beams and columns usually hidden within the walls. Walls are plastered and painted. The upstairs floor is a concrete slab. The roofing is corrugated steel over a lightweight steel structure. Between the ceiling and the roof covering, there is a layer of polystyrene insulation. The window frames are made of aluminium and the ground floor is a 100-millimetre slab finished with ceramic tiles. There are no devices to save energy in the house.

The grandparents live in a small apartment building in the city centre. Their apartment is 66 square metres in area, with two bedrooms, a kitchen, a bathroom and a living room. The apartment building has nine floors with four apartments per floor around a set of two elevators. There is no parking space in the building. The grandparents' maid lives on a plot of land further out from the city where her family have self-built four different houses. The maid is a grandmother, and she shares the plot with her three sons, one daughter, 15 grandchildren and many dogs and cats. Each of the four houses belongs to one of her children. The maid lives in the house of her oldest son along with his wife and their two children. They used to have a patio where they could plant vegetables but recently they have poured concrete over this.

Food

In Tucumán, people used to have four meals: breakfast, lunch, afternoon tea and dinner, with lunch and dinner being the main meals. People used to cook at lunchtime and eat the leftovers for dinner. Following the Spanish and French traditions, breakfast and afternoon tea were light. In between meals, people tended to drink *mate* (a very concentrated herbal tea) with snacks. This has been the routine at the grandparents' house for many years. However, the routine of the new generation is changing little by little. Valentina and Joaquin have breakfast at work around 8 am. They normally have a cup of tea with milk and two *tortillas* (small breads made with animal fat). Around 10 am, Joaquin has fruit and Valentina a cookie. They have a lunch break at 1:30 pm but they go to the grandparents' house to have lunch, which is normally chicken or beef (grilled, schnitzel or roasted) with different salads. It can also be stuffed squash, meatloaf or *esfiha* (Lebanese open-faced meat pies, a bit like a small pizza), depending on the mood of the grandparents' maid. After lunch, they come back to work but the grandparents take a siesta until 4 pm. Valentina and Joaquin keep working in the office until 5 or 6 pm.

Dinner at home is normally around 9:30 pm for everybody, including the baby, and is generally lighter than lunch. It could be vegetable soup with beef, a chicken pie or even leftovers from lunch. On Fridays and Saturdays, dinner can be heavier because it is the day for making *asados* (barbecues). For a barbecue, the person in charge will provide 500 grams of meat for each man and 300 grams for each woman. While the meat is being grilled, people normally make a *picada* (table of cold meats and cheese). The first plate is normally *chorizo* (sausage) with salad. The main plate is usually beef or pork with Russian salad (potato, carrot, peas and mayonnaise). People will drink wine, beer or Fernet (a bitter herbal liqueur) with cola.

On other Saturdays and Sundays, the family usually go out together for dinner and sometimes for lunch. Once or twice a month they receive the grandparents for a meal that might be a barbecue or something special.

In the house of the grandparents' maid, each family eats lunch and dinner independently but the food is very similar to that described above, with the difference that cheaper cuts of meat are used instead of premium ones and pasta is regularly on the menu.

Consumer goods

At home, Valentina and Joaquin have three TVs (32 inches, 30 inches and 27 inches) that were bought 2 years ago when they moved to the new house. The kitchen is fully equipped with a dishwasher, washing machine, dryer, oven, one fridge with a freezer and an extra freezer for barbecues. They also have two laptops and a desktop computer. For cooling and heating the house, they have five air conditioners (one of 8 kW and four of 6kW) and two heaters. The house has an alarm system and two lights that are turned on for the whole night for safety. Inside the house, they also have security cameras in different rooms.

The grandparents have two TVs (30 inches and 27 inches), a washing machine, oven, fridge, microwave and only one laptop computer. They also have two air conditioners, one in the living room (8 kW) and one in the main bedroom.

Energy

In Argentina, the production of electricity is 60 per cent thermal (powered by natural gas), 33 per cent hydroelectric, 5 per cent nuclear, less than 1 per cent wind and solar and the remaining percentage is imported. The electricity consumed in Tucumán is produced by three hydroelectric and four thermal power plants. The household's electricity consumption varies a lot between summer and autumn. In summer, the consumption ranges from 1,000 to 1,200 kilowatt hours per month while in autumn it could be around 500 kilowatt hours. The consumption of electricity is high considering that 90 per cent of people in Tucumán on average consume less than 500 kilowatt hours. The consumption of natural gas in winter is 74 cubic metres per month (800 kilowatt hours/month) and around 30 cubic metres (320 kilowatt hours/month) in summer. The average consumption of natural gas in Tucumán is 40 cubic metres per month (430 kilowatt hours/month) (conversion factors for gas from Packer, 2011). In both households, the maids' rooms are mainly cooled using fans. The consumption of energy for this is unknown, but a fan is not a high electricity consumer. The consumption of water is unknown.

Transportation

There are two cars in the house. Valentina's car is a small SUV-type vehicle. She refills her tank twice per month, which allows her to travel more or less 600 kilometres.

Joaquin's car is a small sedan. Since they both work in the same place in the city, they travel approximately the same distance each month. To go to work, the couple use Valentina's car from Monday to Wednesday and Joaquin's on Thursday and Friday. Valentina uses the bus to return home on Thursday and Friday and Joaquin comes home on the bus on Monday and Wednesday. Even though there are three bicycles in the house, they are hardly used. At the weekends, the couple use their cars independently unless they go out to dine together or visit common friends. The family use the car to go on vacation once a year to the beach in Argentina (1,500 kilometres each way from Tucumán; Argentina is a large country) or they go overseas.

The grandparents have one small car. It is usually parked two blocks away from home because apartment buildings constructed during the 1960s and 1970s, like theirs, rarely had parking spaces. The car is only used during the weekends to visit their business outside the city centre (14 kilometres), for visiting her daughter (once or twice a month) and for dining out. Therefore, they drive a maximum of 40 kilometres per week. They walk or use the bus to go anywhere else in the city, either for work or leisure. Grandfather walks to work every day. Grandmother goes to work outside the city three times a week by bus. They used to take very short vacations to a place 70 kilometres away from the city three times a year.

The family's maid (Lucia) sleeps in the house from Monday to Friday and returns to her own house for the weekends. She lives 40 kilometres away, which means a 2-hour bus ride. During the week, Lucia mainly stays in the house, with a trip to the supermarket with Valentina to take care of the baby while shopping. The grandparents' maid lives 3 kilometres from the city centre which allows her to commute to their house daily by bus.

Conclusion

In Tucumán, the definition of family boundaries has to be approached carefully because people can live in or use different houses on the same day on a regular basis and the consumption and flow of resources within a family are not limited to one house. If the total food consumption of the grandparents' house is measured and divided by two, the result will not describe the real consumption per capita of the house since resources are shared daily with other family members. Moreover, it is important to recognise that changes in the routine of one family might affect other family members outside that group. In the case analysed, the choice of the family to live outside the city centre has had an impact on the routine of the grandparents and the work of their maid, who has to cook for more people.

The tendency of middle-class families to live outside the traditional centre of Tucumán is making everyday life more complicated for everybody. The increasing car dependency has altered the traffic in the city, limiting the mobility of family members and pushing people to buy more cars. This is a big problem for a city that has streets 8–11 metres wide within the traditional core, where all the administrative offices and main shops are still located. Moreover, the new generation of middle-class families living in gated communities has a "higher" lifestyle than previous generations which also means more consumption of goods and resources.

Pedro and Gabriele: Soy and beef farming: Brazil

Writer and researcher: Fabricio Chicca

The family and their farms

This is a Brazilian family of five members—husband (Pedro) and wife (Gabriele), 15-year-old twin boys (Victor and Luiz) and a 13-year-old girl (Julia). The family live on the urban fringe of the city of Dourados in the state of Mato Grosso do Sul in the Brazilian middle west. The family are considered wealthy when compared with the Brazilian standard. Mato Grosso do Sul is a landlocked state with an international border with Paraguay to the west. Farming beef cattle and growing soy and sugar cane account for the most important products of the region. Pedro runs two properties; the one producing soy is 180 kilometres from his house, and the smaller farm, which produces beef cattle, is where the family live. The soy farm is a 1,200 hectare property, making it one of the largest producers in the region. Lately, a small portion of the farm (around 60 hectares) has been dedicated to the production of sugar cane, and a potential shift from soy to cane is being analysed.

Vehicles

The family house is located 21 kilometres from the city centre, with 18 kilometres of tarmac road and 3 kilometres of a private gravel road. The family have eight vehicles, of which four are mostly dedicated to work, two are used for the family routine and two for leisure.

The house

The family live in a very large house, 785 square metres in area plus a veranda around the house. It is divided into seven bedrooms, three living rooms, an office, a meeting room, a dining room, a kitchen, two further bedrooms for maids, five bathrooms and an external area for events and parties. This party area is a large patio, with a 64-square metre covered open area and a 50-square metre support area with kitchen, toilets and storage and a fully equipped gym. This area is located beside a large swimming pool (25 metres by 12 metres) and most reunions of family and friends happen in this area. The house has a concrete structure, with hollow bricks covered by plaster and paint. The roof is the traditional hipped roof, with a 4-metre wide veranda around the entire house. Except for the living rooms, the entire house has a floor finish of concrete slabs. The living room has the roof trusses exposed to create more volume as an aid to natural ventilation. The bedrooms have windows and also French windows opening on to the veranda. All bedrooms have air-conditioning for the summer days. Despite the air-conditioning, the house has very high ceilings and four manually controlled high-level apertures that can be opened to vent the hot air. The bedrooms, in addition, have good ventilation options and even during hot nights the air conditioning systems are rarely used. The family have two televisions, all the usual major kitchen appliances and two 500-litre freezers, five laptops, two desktops (for the office), a stereo in the main living room, two other portable stereos that can be moved around the house and all rooms have ceiling fans. In the month before the time described, the family consumed 444 kilowatt hours of electricity, which is consistent with the family average for the year.

Water is provided through two wells and a rainwater collection system that has been added to provide water for gardening and cleaning. There are also two septic tanks for sewage treatment.

Five people work to maintain the house, a cook, two cleaners, a driver and a general manager, who also works as a secretary for the farm business. There are ten other employees who work on the farm and occasionally also help with some housework, such as gardening, and with parties and events (often barbecues).

Routine

The family day starts at 6 am, when the children have their breakfast. This is normally a traditional Brazilian upper-class breakfast with bread (baked on the farm), jam, butter, cheese, ham, salami, a fruit salad, orange juice and milk (from the farm) with chocolate. They leave for school around 6:30 am with the family driver. The school is 25 kilometres from home and the distance is covered in around 35 minutes. On Wednesdays, Pedro drives the kids to the school as every Wednesday he goes to the soy farm (around 3 hours' drive from home). He normally gets back by the end of the day, but during the week described, he slept over at the soy farm. The parents' day starts around 8 am. Gabriele usually has granola and yoghurt (produced on the farm) and black coffee, while Pedro only eats bread with butter (both made on the farm) with a glass of black coffee. After breakfast, Pedro talks to the soy farm by radio for a daily report. This is normally a brief contact and, except when something very serious has happened the day before, it is just a routine radio talk. After that, there are some vital tasks to be done. The first is to check the international soy price. The next is to review the weather forecast for the region and for the other major producer areas around the world. Pedro also looks for international issues that may affect the cost of soy, a process he calls market evaluation. This all happens in his office at home and it takes him around 2 hours each day. These 2 hours in his office after breakfast are the most consistent part of his week, along with the trip to the soy farm on Wednesdays. The rest of the week's routine is very flexible and depends on the demands made by the farms. However, Pedro likes to go to the city in the afternoons at least every other day. The farm headquarters is officially in downtown Dourados (20 kilometres from the farm). It is a small office where an accountant and secretary work running the accounting and other bureaucratic issues related to the farms. Pedro has a small room and there is also a small meeting room.

Gabriele has a more flexible routine. She goes to town almost every day during the week. She does not have a formal job, but often goes to her sister's fashion shop in town (22 kilometres away) and helps her during the day. She is now studying English and taking private classes in town every day except for Fridays.

The routine for the children is based on their afternoon activities. On Mondays, Wednesdays and Fridays they do not come back home for lunch but instead have lunch in town in a restaurant near the school. At 2 pm, the driver picks them up and takes them to their extra-curricular activities. First, the driver leaves Julia at her dance classes (1 kilometre away) and afterwards she walks (300 metres) to the English classes where she meets her siblings as they all take English classes in the same place. Before the classes, Victor and Luiz go to the gym for an hour (beside the school) where they play tennis, practise judo and exercise. After that, the driver brings them to the English classes (1.3 kilometres away). The driver waits for then until 4:30 pm when he drives them home. On Tuesdays and Thursdays, the driver picks the children up from school and drives them home for lunch.

Food and travel

During the week described, on Monday both Gabriele and Pedro had their lunch at the farm. They had a traditional regional dish, *arroz de carreteiro*, which is rice and dried red meat (normally left over from a barbecue), and a lettuce salad, with orange juice to drink. For dessert, they had papaya. In the afternoon, both went to town in different cars. They drove 55 kilometres and 48 kilometres, respectively. On Monday the children had their routine as described; the driver drove a total of 95 kilometres as he also brought some groceries home. The children always eat in a buffet restaurant in town. Victor and Julia had pasta with bolognese sauce, while Luiz had rice, mashed potato and fried fish. They did not have a dessert, but each had a Coca-Cola. For dinner, the family had the leftovers from lunch, except for Gabriele who had a cucumber and tomato salad with feta cheese.

On Tuesday, Pedro worked from home, so he and Gabriele did not go to town and the children were picked up by the driver. They had their normal breakfast and for lunch, the family had rice and black beans, a lettuce and tomato salad and *porpetta* (meatballs with tomato sauce), with a milk pudding and fruit salad for dessert. In the afternoon, Pedro had a tapioca flour pancake (a traditional indigenous dish) with feta cheese. For dinner, some of the family had the leftovers from lunch while Luiz and Pedro had steaks with onion sauce. The driver went to town to bring the children from school and to deliver some papers around the town. He drove 61 kilometres in total.

On Wednesday, the family had their traditional breakfast together, then Pedro went to the soy farm, returning the next day after lunch. The soy farm has a more modest home, a fully furnished two-bedroom house of around 60 square metres, with all basic appliances. The house is totally powered by solar panels, as there is no electricity grid available. The solar system was installed 3 years ago, and it has been very successful, replacing an old diesel generator. However, all other areas of the farm are powered by diesel.

On Wednesday, for their lunch in town, they all had *feijoada* (black beans cooked with dried meat), which is the most traditional meal in Brazil and normally served only on Wednesdays and Saturdays. It is served with rice and fried cabbage. Gabriele skipped lunch. For dinner, the family had beans and rice and polenta with tomato sauce, orange juice and ice cream for dessert. Pedro ate with the farm employees. For lunch and dinner, they had rice and beans, manioc and a meat stew.

On Wednesday, the driver drove 88 kilometres as he also stopped at the supermarket. Gabriele went to English classes and stopped for a couple of hours at her sister's shop; for this, she drove 58 kilometres.

On Thursday, the family had their normal breakfast, except Gabriele, who had a tapioca pancake with coconut milk sauce. At the soy farm, for breakfast, Pedro had a couple of apples and coffee. On his way back, he stopped by the road to have a traditional Paraguay snack called *chipa*, a kind of bread made of manioc flour. He drove a total of 401 kilometres over Wednesday and Thursday. The family had rice, black beans, grilled chicken breast and green salad for lunch. Pedro ate when he got home in the middle of the afternoon and had the leftovers from lunch, after which he did not leave home. For dinner, the family had rice and beans, fried cabbage, lettuce and tomato salad and chicken pie. They had ice cream for dessert and orange and grape juice. The driver went 68 kilometres and Gabriele drove 51 kilometres as she went to town for English classes and the visit to her sister.

On Friday, the family had their breakfasts at their normal times. Pedro had two scrambled eggs in addition to his bread. Gabriele went to town just after 10 am to meet some friends, and to buy groceries for a barbecue on Saturday afternoon. She had lunch with the children in the same restaurant. That day, they had a very mixed selection of food. Gabriele had a chicken breast with green salad, Julia had a cheeseburger and salad, Luiz had steak with mushroom sauce and Victor had stroganoff served with fried chips and rice. They all had ice cream for dessert, except Gabriele who had fruit salad and a coffee. At home, Pedro had pasta with tomato sauce and a steak. On Friday night, the family went to town for dinner with friends. They had six large pizzas, three 600-millilitre beers and a *caipirinha* (a traditional Brazilian cocktail), followed by fruit salad for the mother and daughter, pudding for the father and ice cream for the boys. Gabriele drove 77 kilometres, while the driver drove 65 kilometres plus a further 42 kilometres for the dinner round trip.

At the weekends, the family have a more relaxed schedule. The breakfast table is set around 7 am and it is not cleared until around 11 am or when the last family member wakes (normally the boys). The breakfast is much the same as for the rest of the week. Despite it being the weekend, Pedro had his breakfast at his traditional time, and had his daily 2-hour routine in his office assessing the market and talking to the soy farm. Gabriele had her breakfast at 9:30 am and went to town with the cook and one of the house cleaners for the weekly groceries shop. They got back at 12:30 pm and then they started the preparations for the afternoon barbecue, having invited two families to come over. In total, there were 18 people present at the barbecue. At the event, they consumed 14 kilograms of red meat (from the farm), 5 kilograms of manioc (in that part of the country, a barbecue is traditionally served with manioc), 3 cups of rice, a litre of vinaigrette dressing, 21 600-millilitre bottles of beer, 12 litres of soft drinks, 8 litres of ice cream, 2 papayas, 12 oranges and 14 small baguettes. The barbecue started around 2 pm and stretched on until 10 pm, when one family left and the other decided to spend the night there. Gabriele drove 71 kilometres to town and back.

On Sunday, the same arrangement was made for breakfast. The family and their five guests had the same traditional breakfast with the addition of eggs (14) and bacon (2.2 kilograms), which Pedro and Gabriele did not eat. Pedro woke at the normal time for his office routine. For lunch, the family and guests had *arroz de carreteiro* made with the leftovers from the barbecue, with the addition of a green salad and manioc (2 kilograms), and with papayas (6) and mangoes (10) for dessert. For dinner, the family had the remaining *arroz de carreteiro* while the boys had cheese and ham sandwiches. They had chocolate ice cream for dessert.

Felipe and Laura: Living with car restrictions: Brazil

Writer and researcher: Fabricio Chicca

The family, their apartment and vehicles

The family consist of Felipe and Laura, their daughter aged 13 and son aged 10. They live in São Paulo, the largest city in Brazil, in an apartment on the thirteenth floor of a 24-storey building in Pinheiros, a high-class neighbourhood. The apartment area is 101 square metres and it comes with a parking space for two cars. The building has a concrete structure with partitions of plastered and painted hollow-fired brick.

They have two modest cars, both "Flex" models which can be fuelled either with gasoline or with sugar cane ethanol. This happens according to the seasons, as the price of ethanol changes over the year. If the ethanol is not 30 per cent cheaper, the family use gasoline, as it is the more efficient fuel. The family need two cars despite only using one during the week. This is because São Paulo imposes a car circulation restriction once a week according to the car's registration plate. As they drive only one car to work, and leisure activities happen inside a small area, they often manage to go through the month with only one tank of fuel.

The apartment has three bedrooms and a very small office, which is used as a study room for the children. There are three televisions (one for each child in his/her room and one in the living room) and the kitchen has all the standard appliances. The apartment has air-conditioning; however, even during the hot months of the year, the family barely use this. Felipe is really into technology, and as a result, the family's electronic devices are often updated. The couple each have a laptop and the children have their own tablets. The family consume, on average, 800 kilowatt hours per month. In July 2016, the consumption was 1,052 kilowatt hours, a bit higher than normal as they had guests staying over for 3 weeks. For the same period, they used 4 cubic metres of gas (40 kilowatt hours), which matches their average consumption.

Weekday routine and diet

The family believe that they are very organised and an inflexible schedule is necessary to make the family operate effectively. Twice a week, on Mondays and Wednesdays, Felipe is responsible for overseeing the children's breakfast. They normally have a slice of bread with butter and jam, fruit and a glass of milk. The parents have just fruit and a glass of coffee and some milk. Occasionally Felipe has bread, but normally only fruit with a black coffee. Laura often skips breakfast. On Tuesdays, before she goes to work, she walks three blocks to a vegetable market, where she buys most of the produce for the week. On Felipe's breakfast days, she goes down to the ground floor to exercise in the building's gym. The family leave the house around 7:45 am in one of the cars, first to drop the children at their school (3.4 kilometres from home) and then Felipe at his work (a further 4.5 kilometres). Laura then drives to her work (a further 2.8 kilometres). The parents occasionally leave their work for meetings and when they do, they do not use their own car.

The children have their lunch at school, together with snacks in the middle of the morning and afternoon. The school is a traditional Brazilian school, so serves traditional Brazilian food, which is basically rice and beans, with a meat option, normally red meat or chicken, and a salad. On Fridays, fish is served, and at least once a week a pasta dish is offered instead of the traditional rice and beans. Both Felipe and Laura always have lunch out, as their companies provide them with food vouchers (a very common practice in Brazil). Felipe has a voucher of US$12/day while Laura has US$19/day (August 2016). Their first preference is also Brazilian food, but Laura said she is reducing the amount of red meat she eats, aiming to have it only once a week. Felipe, however, believes he has what would be described as the traditional diet, always rice and beans and meat, and sometimes salad. Despite having the financial means to have more sophisticated food, he still chooses the most traditional. Felipe is an engineer and Laura an accountant. The family have a house cleaner who helps with the cleaning twice a week, on Tuesdays and Thursdays. Those days, the children take the school bus home and the cleaner looks after them until the couple (or at least one of them) come home. They normally come back from work together, but Laura occasionally has to work longer hours, and on those days Felipe comes back home by bus. This happens at least twice a month. On Mondays and Wednesdays, the children go to their grandparents' house, which is conveniently located across the street from the school. At the end of the working day, they are picked up by their parents. On Fridays, the children have extra activities at the school and then their parents pick them up after work.

According to the Brazilian tradition, the family have a lighter meal for dinner. They normally have sandwiches, salad and some fruit. During the observation period, the family had a pizza on Thursday, which is unusual.

At the weekends, the family have an arrangement regarding entertainment. The children are allowed to pick one activity for them and the parents have a grown-up time for themselves. According to Felipe, the arrangement has worked well since they created it 5 years ago. The children normally like to go to the shopping mall, where many activities are available, such as an amusement park, cinemas and playground. The parents, on the other hand, like to go out at night, either on Friday or Saturday. The most common leisure activity for them would be to go out for dinner and movies. Lately, they have gone to concerts as Felipe is also a musician. The vast majority of these activities are available within a 5-kilometre radius from their apartment but they sometimes have problems over agreeing on things to do together. The meals during the weekend are similar to those during the week, with more time flexibility. On Sunday, however, the family normally choose Italian food for lunch (both parents are from Italian families in their third generation in Brazil). They try to organise buying the groceries during the weekend, to avoid the hassle of doing it during the week. They drive to a supermarket 4.2 kilometres from home, normally either on Saturday morning or Sunday night. During the week, they go to a bakery located at the corner of their street. Besides bread, they also buy the ingredients for filling the night's sandwiches and, when necessary, they go to a small local shop to buy small goods, although they try to avoid this as the prices are considerably higher than at the supermarket. During the week of observation, the family bought bread every night, and cheese and ham (or similar) on Mondays and Wednesday.

During the weekend there is no car circulation restriction in São Paulo, and the family always choose the more comfortable car. During the observation weekend, the family drove 24 kilometres.

References

Brazilian Travel Information (2015) São Paulo history, available at http://www.brazil travelinformation.com/brazil_saopaulo_history.htm, accessed 6 July 2017.

Climate-Data (n.d.) *Climate: Dourados*, available at https://en.climate-data.org/location/314747/, accessed 6 July 2017.

Countries of the World (2017) List of countries in South America, available from https://www.countries-ofthe-world.com/countries-of-south-america.html, accessed 3 July 2017.

Dourados Prefeitura (2017a) Historical synthesis, available at http://www.dourados.ms.gov.br/, accessed 6 July 2017.

Dourados Prefeitura (2017b) City of Dourados, available at http://www.dourados.ms.gov.br/index.php/cidade-de-dourados/, accessed 6 July 2017.

Earth Observatory (2014) *Growth of São Paulo, Brazil*, available at https://earthobservatory.nasa.gov/IOTD/view.php?id=83947, accessed 6 July 2017.

Encyclopaedia Britannica (n.d.) *Tucumán*, available at https://www.britannica.com/place/Tucumán, accessed 5 July 2017.

Encyclopaedia Britannica (2017) São Paulo, Brazil, available at https://www.britannica.com/place/Sao-Paulo-Brazil/Climate, accessed 6 July 2017.

Global Footprint Network (GFN) (2016) *National Footprint Accounts 2016 Edition*, available from https://data.world/footprint/nfa-2017-edition, accessed 28 June 2017.

Global Footprint Network (GFN) (2017) Ecological wealth of nations, available at http://www.footprintnetwork.org/content/documents/ecological_footprint_nations/biocapacity_per_capita.html, accessed 4 July 2017.

Index mundi (n.d.) *Population below poverty line – South America*, available at http://www.indexmundi.com/map/?v=69&r=sa&l=en, accessed 4 July 2017.

Knoema (2017) *Tucumán – Total population*, available at https://knoema.com/atlas/Argentina/Tucum%c3%a1n/Total-Population, accessed 6 July 2017.

Malizia M (2015) The study of gated communities: A methodological proposal applied to the Yerba Buena municipality (Gran San Miguel de Tucumán agglomerate in the Northwest of Argentina), *Estudios demográficos y urbanos*, 30(1), pp. 103–134.

Neto S, Duarte G and Páez A (2015) Gender and commuting time in São Paulo Metropolitan Region, *Urban Studies*, 52(2), pp. 298–313.

NOSSASP (2014) Rede Nossa São Paulo e Ibope lançam oitava pesquisa sobre Mobilidade Urbana http://www.nossasaopaulo.org.br/noticias/rede-nossa-sao-paulo-e-ibope-lancam-oitava-pesquisa-sobre-mobilidade-urbana, accessed 6 July 2017.

Packer N (2011) *A beginner's guide to energy and power*. Presented as part of the Renewable Energies Transfer System Project (RETS) funded by INTERREG IVC through the European Regional Development Fund. Stoke-on-Trent, Staffordshire University, February 2011, available at http://www.rets-project.eu/UserFiles/File/pdf/respedia/A-Beginners-Guide-to-Energy-and-Power-EN.pdf, accessed 7 July 2017.

Perrotta-Bosch (2016) Dissatisfied São Paulo, *Architectural Design*, 86(3), pp. 60–69.

PopulationLabs (2011) Argentina population map, available at http://www.populationlabs.com/argentina_population.asp, accessed 6 July 2017.

Schiffman R (2015) Rain-forest threats resume, *Scientific American*, 321(6), p.24.

Tavares M (2015) A Toponímia das Localidades Rurais do Município de Dourados/MS, *Revista do GEL*, 12(2), pp. 164–191.

United Press International NewsTrack (2007) *Scientists name Amazon the longest river*, available at http://go.galegroup.com/ps/i.do?&id=GALE|A165120537&v=2.1&u=vuw&it=r&p=ITOF&sw=w&authCount=1, accessed 5 July 2017.

What Are the 7 Continents (2017) *The 7 continents of the world*, available at http://www.whatarethe7continents.com/south-america/, accessed 3 July 2017.

Zimmermann K A (2013) *Andes: World's longest mountain range*, available at https://www.livescience.com/27897-andes-mountains.html, accessed 3 July 2017.

8
CALCULATING THE ECOLOGICAL FOOTPRINTS OF THE STORIES

Fabricio Chicca, Sarah Nabyl-Calliou, Robert Vale and Brenda Vale

Introduction

The calculation of environmental impact does not represent the total impact of the families in this research. We tried to minimise the assumptions necessary to produce a quantifiable figure for the environmental impact caused by each family. However, when some "educated assumptions" had to be made, we chose the most environmentally favourable number available, except when we explicitly state otherwise. We wanted to be sure that we were not making things out to be worse than they might be. Additionally, we avoided making these kinds of assumptions for any items where a change would represent a major change in the results. For such items, additional data were collected and "equalised", by which we mean that we used a number of sources and checked the origin of the data, and whether the data is measuring the same stuff.

Methodology

To calculate the environmental impact means that we have to identify the specificities that apply to each country. The simple idea of comparing how much land is necessary to provide the needs of the families in the stories and to sequester their CO_2 emissions is more complicated than it seems. Land in different parts of the world has different yields for crops and different capacities to absorb CO_2. A hectare of New Zealand's productive pastureland has a very different productivity to a hectare in Antarctica. The comparison between a hectare from Antarctica and one in New Zealand would be environmentally meaningless. To deal with this, the Global Footprint Network (GFN), based on the concept devised by Wackernagel and Rees (1996), uses the notion of a "global hectare", which is a hectare of land with the global average productivity. The concept of global hectare is based on the concept of biocapacity. The GFN defines ecological footprint (EF) and biocapacity as follows:

> On the demand side, the Ecological Footprint measures a country's use of cropland, forests, grazing land, and fishing grounds for providing resources and absorbing carbon dioxide from burning fossil fuels. On the supply side, biocapacity measures how much biologically productive area is available to regenerate these resources and services.
>
> *(Global Footprint Network, 2017)*

The global hectare allows researchers to account for the available biocapacity and the demand on biocapacity (the EF), which we all make by our demands for goods and services. The total biocapacity for a whole country can be obtained by multiplying the biocapacity value published by the GFN in global hectares per person by the total population of the country. Although we have used the concept and some data from the GFN, our calculations for the specific impact follow a similar but slightly different methodology. Knowing the total biocapacity for each country has allowed us to calculate a "country index" for converting hectares of land in a given country into global hectares by dividing the total biocapacity in global hectares by the area of the country in "normal" hectares. Using the biocapacity for a particular country takes into account a number of factors, including the amount of land taken up by buildings and infrastructure, the capacity to generate energy or to sequester carbon dioxide, and how fertile the soil is.

Knowing the total biocapacity, we then need to assess the total amount of land allocated to cropland, grazing land, forest land, fishing grounds and built-up land. Each of those items is defined by GFN. We followed these definitions to separate each item consumed by the families into categories and allocated the items to the appropriate land types. When the amount of land was calculated (by yields for food, or CO_2 sequestration for energy) for any specific item, we multiplied it by our country index, which considers the relationship between real hectares and global hectares for each country, to obtain a value in global hectares.

We have also produced a calculation for CO_2 emissions, for which the methodology is presented after the discussion about EF.

Calculation for food (EF and CO_2)

Food was the most complex item for which to calculate the environmental impact. We have dedicated part of the food chapter (Chapter 10) to talking further about factors that affect the environmental impact from food.

Countries' specificities regarding their agricultural production posed a challenge to making it possible to compare their environmental impact. The yield for each agricultural item is often calculated differently for each country's statistics. For this reason, we decided to use data for food from the United Nations Food and Agriculture Organisation (FAO). This provides figures for population, production, land use and agricultural CO_2 emissions, as well as a useful factor called the "Commodity Balance", which is discussed below. From the FAO yield data for a particular country, we were then able to convert the data from hectares to global hectares by applying that country's index for the item according to its origin (whether it was produced from cropland or grazing land).

Following the division used by GFN, we divided the agricultural products into two categories:

- Livestock products: meat, milk, butter, eggs, etc.
- Crop products: fruits, vegetables, cereals, etc.

The calculation of the ecological footprint is essentially the same for both; the only difference is the calculation of the yield. The calculation is as follows:

Knowing the family's consumption of each item in kilograms, to calculate the footprint in global hectares, we need to calculate the yield in kilograms/global hectare.

Inputs are:

- The family's *consumption* for every item in *kilograms* by item of food → c_i
- The country's *yield* per item of *crop products* in *hectograms*/hectare* for the year 2014 → y_i
- The country's *production* per item of *livestock products* in *tonnes* for the year 2014 → p_i

*Hg is a hectogram, which is 100 grams, so there are 10 hectograms in a kilogram. This uncommon unit is the one used by FAO for their yield data.
This difference between crop products and livestock products arises because the available data from FAO for livestock products differ from those of crops.

Crop products calculation

The calculation of the ecological footprint (*EF*) for crop products is fairly simple: from the consumption (*kilograms*) and the yield (hectograms/hectare converted to *kilograms/global hectare* with the country's index), we divide the consumption by the yield.

$$EF_i = \frac{c_i}{y_i}$$

Livestock products calculation

The calculation of the ecological footprint for livestock products is less straightforward: the yield figures available on the FAO website for production in each country give no direct relation between the total quantity of that product and the area of land used to produce it. To solve this issue, we used the FAO data in combination with the *Livestock Unit* (LSU), which is a "reference unit which facilitates the aggregation of livestock from various species ... via the use of specific coefficients established initially on the basis of the nutritional or feed requirement of each type of animal" (Eurostat, 2017). One LSU is the grazing equivalent of "one adult dairy cow producing 3 000 kg of milk annually, without additional concentrated foodstuffs" (Eurostat, 2017).

In addition, our calculation has to take into consideration the amount of cropland utilised to feed animals. While the cow, which represents a Livestock Unit, is eating only grass, most cows get supplemental food, as do hens, pigs, sheep and all the others. To take account of this, to the amount of grazing land necessary for livestock has been added the area of land needed to grow the crop products used to feed livestock. The factor called "Commodity Balance", provided by the FAO for each country, shows how much domestic crop supply is available each year (in tonnes) and how much is used to feed livestock. We then have a percentage of crop products used to produce livestock products.

Therefore, l_i, the calculated land area used for a livestock product i (suppose it is beef in Brazil, for example), is the total grazing land in Brazil l_G, which is provided by the GFN, plus the total cropland l_C in Brazil, also from the GFN, multiplied by the percentage of cropland P used for feeding livestock which is given by the FAO Commodity Balance data for Brazil. This sum is then multiplied by the LSU for beef cattle in Brazil:

$$l_i = LSU \cdot l_G + P \cdot l_C$$

We get the yield product, y_i (in *kilograms/global hectare*), by dividing the total annual production p_i (kilograms), using in this example the FAO data for beef produced in Brazil in a year, by the calculated land used l_i (global hectare):

$$y_i = \frac{p_i}{l_i}$$

The EF is then calculated using the same method as that for crop products:

$$EF_i = \frac{c_i}{y_i}$$

The matter of fast food: the Burger Converter

The impact of fast food was considered as a special case. We determined that fast food has 4.78 times the environmental impact of similar food consumed at home (Vale & Vale, 2009, p. 42). We implemented an indicator, which we call the Burger Converter, that gives an average footprint for a fast-food meal that we could multiply by the number of meals eaten out. We considered 100 grams of meat for a burger and 30 grams of cheese. Initially, we also considered fries and a fizzy drink, but we decided the impact of these was irrelevant as the meat and cheese would have 42 times more impact than these items.

The sum of all footprints per item divided by the number of people in the family n gives the final footprint for food in global hectares per person:

$$EF = \frac{\sum_i EF_i}{n}$$

CO_2 and CO_{2eq} emissions for food

In some cases, as for food, for instance, the combination of the two methods, EF and CO_2 emissions, was extremely helpful for a fuller understanding of the environmental impact. As the FAO provides reliable online data regarding CO_2 emissions for agricultural items per country, it was relatively simple to calculate the emissions for food.

Calculation for travel (CO_2 and EF)

Each family had their assessment made based on the transportation and its CO_2 emissions, including the multiplier for the embodied energy for extraction, refining and shipping of petroleum products of 1.25. From the amount of CO_2 produced by the total of transportation modes from each family, we transformed the emissions into hectares, considering the sequestration capacity according to the average for each climate as shown in Table 8.1.

The formula to calculate the EF for travel is presented below:

- Climate sequestration capacity (Table 8.1—IPCC 2007) in *tonnes/hectare/year* → c_i
- Family travel emissions in *kilograms* → ft_e
- Land for CO_2 sequestration in *hectares* → L

The equation is:

$$L = \frac{ft_e}{c_i}$$

TABLE 8.1 Carbon dioxide sequestration potential in three climates from IPCC (2007)

Climate	Capacity of CO_2 sequestration in tonnes/hectare/year
Tropical	6.0
Moderate	3.0
Boreal	0.8

Calculating the ecological footprints of the stories **131**

The result in hectares is then multiplied by the country index C_i and divided by the number of people:

$$EF = \frac{L \star C_i}{n}$$

The application of the country index aims to include the impact of regional infrastructure in the calculation. The index directly reflects the amount of infrastructure and services per capita and how much productive land has been extracted from the environment to become infrastructure.

Calculation for energy (EF and CO_2)

EF of electricity

The energy calculation began with determining the impact of the electricity used by the families in the stories. This was based on data from Chambers et al. (2000: 82–83) which determines the average environmental impact from different sources of energy used for electricity generation (see Table 8.2). We used these footprint factors for a number of renewable and non-renewable sources of electricity in *hectare years/gigawatt hour*.

From the electricity generation energy mix for each country, we calculated the footprint for each energy source i.

Inputs:

- Family's consumption per year (*gigawatt hours/year*) → C
- Energy source's proportion in energy mix (*per cent*) → p_i
- Energy source's footprint (*hectare years per gigawatt hour*) → EF_T

$$EF_i = C.p_i .EF_T$$

The sum of all footprints per energy source divided by the number of people in the family n is the final footprint for electricity:

$$EF = \frac{\sum_i EF_i}{n}$$

TABLE 8.2 Ecological footprints for a range of electricity generation options

Energy: Electricity generation	Coal	Oil	Natural gas	Wind	Photovoltaics	Biomass	Hydroelectric
Footprint (ha years per GWh)	198	150	94	6	24	36.5	42.5

TABLE 8.3 Coal equivalents of fuels used directly (European Nuclear Society, 2017)

1 kg gasoline	1.59 kg coal equivalent
1 kg fuel oil	1.52 kg coal equivalent
1 m³ natural gas	1.35 kg coal equivalent
1 kg hard coal	1.00 kg coal equivalent
1 m³ town gas	0.60 kg coal equivalent
1 kg firewood	0.57 kg coal equivalent
1 kg crude lignite	0.34 kg coal equivalent

Other energy sources

Some specific fuels such as wood, coal, oil or gas were utilised by some households in addition to electricity. A quantity in kilograms or cubic metres of a fuel can be converted using coal equivalents as shown in Table 8.3. The European Nuclear Society has suggested that a kilogram of coal has a calorific value of 8,141 kilowatt hours/kilogram. We then can use this value with the coal equivalent to calculate the energy content for other sources.

We then used these values to calculate an energy consumption in gigawatt hours/year and the calculation is the same as above, without the energy mix proportions.

$$EF_i = C \cdot EF_T$$

The sum of all footprints per energy source divided by the number of people in the family n is the final footprint for energy other than electricity:

$$EF = \frac{\sum_i EF_i}{n}$$

The total footprint for energy is the addition of all energy footprints.

CO₂ emissions for domestic electricity

The calculation of CO_2 emissions for domestic use was based on the electricity usage and the energy mix from each country. Except where information was provided, as for the Germany family, the emissions were calculated from the electricity generation energy mix for the country. For instance, in Table 8.4 is the energy mix for Finland (International Energy Agency, 2013).

Considering the total consumption of the Finnish family—0.010794 gigawatt hours/year—the CO_{2eq} was calculated, firstly calculating the amount of energy per source as follows:

Energy source (per cent) → Es_i
Family's consumption per year (gigawatt hours/year) → C
Energy per source (gigawatt hours) → Eps_i

$$Eps_i = C \star Es_i$$

So,

$$Eps_{i\ coal} = 0.010794 \star 14\%$$

then,

$$Eps_{i\ coal} = 0.0015\ Gwh$$

The above calculation is applied to each source. The results will be multiplied by the emissions from each source, as can be seen in Table 8.5.

The following calculation was simply the multiplication of the amount of energy used per source by the emissions from the table above, for example, for coal in Finland:

$$GHG_{coal} = 888 \star 0.015$$

$$GHG_{coal} = 0.14tn$$

TABLE 8.4 Energy mix for the generation of electricity in Finland (International Energy Agency, 2013)

Energy source	Proportion(%)
Coal	14
Wind	0.5
Oil	0.8
Hydroelectricity	16.9
Solar	0.6
Biomass	22.7
Nuclear	31.6
Natural gas	12.9
Imported	0

TABLE 8.5 Average CO_2 emissions for electricity generation sources

Source of electricity	Tonnes CO_2/GWh average*
Coal	888
Natural gas	499
Oil	733
Hydroelectric	26
Nuclear	29
Biofuels and waste	45
Wind	26
Solar photovoltaics	85
Geothermal**	80

*World Nuclear Association (2011)
**See Table 12.3

The total greenhouse gas (GHG) emission for electricity is the sum of all the emissions from all sources.

$$GHG_{total} = \sum GHG_{sources}$$

Calculation for housing (EF and CO_2)

EF of houses

For this calculation, we used figures from Vale and Vale (2009) to calculate the amount of energy required to build the house. To this, we applied the calculation for energy using the country's energy mix. Vale and Vale (2009: 145) suggest the embodied energy for houses as shown in Table 8.6.

Using these figures, we calculated the proportional amount of energy necessary to produce the materials and build the house based on the floor area of the house in each story. Then, knowing the amount of energy necessary to produce the materials and build the house, we applied the same principles used to calculate the impact from energy as described previously.

Although the data from Vale and Vale (2009) were applied for the majority of houses, some families, particularly those that had to build their houses using unconventional materials and those with extremely small dwellings, had their environmental impact for houses calculated separately. The result was often not included in the final number as the environmental impact expressed in global hectares was smaller than 0.01.

CO_2 emissions for housing

The calculations for GHG emissions for houses were essentially a calculation of the proportion of energy used to build the house using Table 8.6. After the amount of energy used to build the house was calculated in gigawatt hours based on the floor area, it was divided by the lifespan of the house, assumed here as 50 years. The result is the annual energy usage in gigawatt hours. After that, the same method used for domestic energy was applied.

TABLE 8.6 Embodied energy of houses

House type	Embodied energy (GJ)
100 m² house, timber frame, brick veneer	264
200 m² house, timber frame, brick veneer	461

Calculation for appliances and consumer goods (EF and CO_2)

Appliances

In order to measure the total embodied energy of household products for each case study, the figures established in Bakshi (2017) are used. Material quantities for the

Calculating the ecological footprints of the stories 135

various appliances investigated here are based on a large overall analysis of many examples of each appliance type which established representative figures for material types and quantities used in the manufacture of all appliances of different types. From this, we were able to calculate the amount of CO_2 equivalent emissions related to the manufacture of the appliances owned by each family. Then, for each country, we transformed the emissions into hectares, considering the sequestration capacity according to the average for each climate determined by the IPCC (2007), as given in Table 8.7.

The emissions calculation uses the following information:

- Climate sequestration capacity index in *tonnes/hectare/year* → c_i
- Family appliances emissions in *kilograms* → f_e
- Land for CO_2 sequestration in *hectares* → L

TABLE 8.7 Carbon dioxide sequestration rates for different climates (IPCC, 2007)

Climate	Capacity of CO_2 sequestration in tonnes/hectare/year
Tropical	6.0
Moderate	3.0
Boreal	0.8

TABLE 8.8 CO_2 emissions calculated for each family's appliances

Argentina	0.34
Australia	0.24
Brazil rural	0.60
Brazil urban	0.28
Canada	0.38
Cuba	0.04
Finland	0.45
Germany	0.16
India	0.00
Indonesia	0.00
Japan	0.17
Malaysia	0.32
Mongolia	0.10
Morocco	0.16
Mozambique	0.00
Myanmar	0.01
New Zealand	0.19
South Africa	0.10
South Sudan (Juba)	0.00
South Sudan (UNMISS)	0.00
Tonga	0.22
United Kingdom	0.31
United States rural	0.19
United States urban	0.30

The equation is:

$$L = \frac{f_e}{c_i}$$

The result in hectares is then multiplied by the country index C_i and divided by the number of people:

$$EF = \frac{L \star C_i}{n}$$

Table 8.8 shows the CO_2 emissions for manufacturing appliances for each family.

Calculation for consumer goods (EF and CO_2)

The environmental impact for consumer goods is rather complicated. The exact calculation would require that all items bought or used by the families were accounted for, and every single item would need to have its origin and emissions compiled. As a result, we are estimating the impact using figures from Field (2011). In Field's investigation, 29 per cent of a typical family's expenditure would be considered consumer goods, with 4 per cent related to appliances. Having calculated CO_2 emissions from appliances, we calculated the impact from consumer goods based on knowing the values for appliances (see Table 8.8) and adapting these proportionally.

References

Bakshi N (2017) *A Life Cycle Analysis of Living Measuring Behaviour and the Impact of Dwelling Rather Than the Dwelling Alone*, Phd Thesis. Wellington: Victoria University of Wellington, 6 September 2017.

Chambers N, Simmons C and Wackernagel M (2000) *Sharing Nature's Interest*, London: Earthscan.

European Nuclear Society (2017) Coal equivalent page, available at https://www.euronuclear.org/info/encyclopedia/coalequivalent.htm, accessed 1 August 2017.

Eurostat (2017) Glossary Eurostat, available at http://ec.europa.eu/eurostat/statistics-explained/index.php/Glossary:Livestock_unit_(LSU), accessed 12 August 2017.

Field C (2011) *The Ecological Footprint of Wellingtonians in the 1950s*, Wellington, NZ: Victoria University of Wellington.

Global Footprint Network (2017) Data Global Footprint Network, available at http://data.footprintnetwork.org/, accessed 18 August 2017.

International Energy Agency (2013) *Energy Policies of IEA Countries – Finland*, Paris: International Energy Agency.

IPCC (2007) *Land use, land change and forestry*, available at http://www.ipcc.ch/ipccreports/sres/land_use/index.php?idp=151, accessed 23 August 2017.

Vale R and Vale B (2009) *Time to Eat the Dog? The Real Guide to Sustainable Living*, London: Thames and Hudson.

Wackernagel M and Rees W (1996) *Our Ecological Footprint – Reducing Human Impact on the Earth*, Gabriola Island, British Columbia: New Society Publishers.

World Nuclear Association (2011) Comparison of lifecycle greenhouse gas emissions of various electricity generation sources, available at http://www.world-nuclear.org/uploadedFiles/org/WNA/Publications/Working_Group_Reports/comparison_of_lifecycle.pdf, accessed 21 August 2017.

9
COMPARING THE FOOTPRINTS

Fabricio Chicca, Robert Vale and Brenda Vale

Introduction

This chapter compares the lifestyles of the families in the stories to see the environmental impact of their daily habits. The calculation of this environmental impact is complex and based on a number of assumptions, meaning that comparisons are valid but that individual results should perhaps not be compared directly with figures calculated by others, except in terms of rankings. The aim in this book is to relate lifestyle and its impact to the country where a family lives. The environmental footprints (EFs) calculated here include most of the factors included in other calculations except for the impact of eating fish, which makes only a small difference and only to some of the footprints.

The problem of the EF and where you live

A significant reduction in the environmental impact of families with a high impact could be achieved through changes in lifestyle, but reductions are only valid if compared with families in the same country. Comparison between countries is somewhat more complicated. The calculation for making possible comparisons of impact between different countries uses an index on a country-by-country basis for converting hectares (ha) into global hectares, as explained in Chapter 8. What the index measures is how many global hectares are needed to replace one hectare of land in a given country. This means it considers the environmental profile of the country, including its buildings and infrastructure, all available unoccupied land, its capacity to generate energy or sequester CO_2, and how fertile the soil is. Because of this index, if a person has an impact that needs a hectare of land in New Zealand, it means that they will be using 1.5 global hectares. The higher the index the higher the global environmental cost of living in that country. This index is key to understanding the comparison between the families in this chapter. Despite the fact that we are only assessing 24 families in 21 countries, we have calculated the index for all the countries for which we could find reliable data (164 countries). Table 9.2 shows the index calculations for the countries in which our stories are located. The assessment of the indices for the top twenty countries reveals some predictable results. For instance, 65 per cent of countries in the top 20 rankings in the index are European. (Figure 9.1)

European countries generally have large populations and a lot of built-up land and infrastructure, which means less land available for growing things. However, Bangladesh and Trinidad and Tobago are in sixth and seventh positions in the top 10. The indices for all countries and the densities were calculated using data from the Food and Agricultural Organization (FAO) for population and country sizes from the World Bank (FAO, 2017; World Bank, 2017). All countries in the top 10, when ranked by index, had a relatively high density. Table 9.1 shows the conversion factor (hectares to global hectares) and density for the top 10 countries in the world. The same lifestyle would have to be multiplied by 5.91 in Demark and 3.64 in the United Kingdom to convert the impact to global hectares, land of average world productivity.

The average density for all 164 countries calculated is 0.55 persons/hectare (p/ha), while that of the top 10 countries is 16.4 persons/hectare. Excluding Bahrain, Singapore and Bangladesh, the three countries with the highest densities, the average is still 2.55 persons/hectare. Grouping the highest and lowest indices for all countries in groups of 30, for the top 30 countries the average density is 6.44 persons/hectare, dropping to 0.40 persons/hectare for the bottom 30 group. This suggests that countries with a highly developed infrastructure and/or a high population density will tend to have a high index. The relationship between a high index, being highly developed and/or having a high density makes sense when

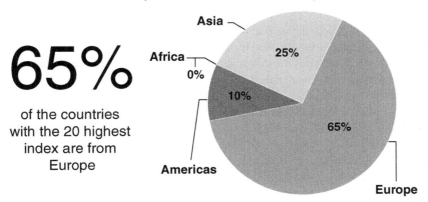

FIGURE 9.1 Location of the countries which are in the Top 20 ranking of the Index.

TABLE 9.1 Conversion factors/indices and densities of countries with the Top 10 indices

Country	Conversion factor, hectares to global hectares	Density persons/hectare
Bahrain	11.60	20.71
Denmark	5.91	1.32
Germany	4.78	2.27
Singapore	4.62	115.49
Netherlands	4.52	4.08
Bangladesh	4.04	10.91
Trinidad and Tobago	4.00	2.65
Belgium	3.96	3.47
United Kingdom	3.64	2.66
Czech Republic	3.45	1.34

the availability of land becomes the main part of the discussion. Highly developed countries normally have a significant portion of their land that would otherwise be contributing to their biocapacity taken up by infrastructure, including buildings, roads, railways, ports, communications, public spaces, and so on.

The quality of the land available for cropland and grazing also influences the conversion factor. A highly developed, dense country with fertile land increases the index, as can be seen for Denmark, Germany and the Netherlands. This is because if the land is more fertile, more global hectares of average fertility will be needed to match one hectare of this more fertile land. Population density will push the factor up for both developed and developing countries. Bangladesh and Singapore, for instance, with national densities of 11 persons/hectare and 115 persons/hectare, respectively, necessarily have a high index as land is scarce. In Bangladesh's case, no matter how low the impact of the people and despite infrastructure being not highly developed, the land is taken up by people.

The top countries in terms of the index tend to have a very high average income, around US$38,000 a year for the top 10. The average for the remaining 154 countries is around US$15,000. When we look at the groups of 30 countries again, apart from the top group there is no clear pattern for the other groups, as can be seen in Figure 9.2.

Average income drops dramatically from the first group of 30 to the other groups. As might be expected, a similar thing happens with the Human Development Index (HDI), although the drop is much less abrupt. The United Nations developed the HDI as an alternative to measuring development by economic growth. It is "a summary measure of average achievement in key dimensions of human development: a long and healthy life, being knowledgeable and ... a decent standard of living" (UNDP, 2016a). An HDI over 0.7 is considered to be high—and over around 0.79 very high—human development (GFN, 2017). As a general rule, a country with a relatively high density (higher than 2 persons/hectare), high HDI and high average income is more likely to have a high index. This can be summarised by saying that high density makes land scarce, while high income and HDI suggest land taken by buildings and infrastructure.

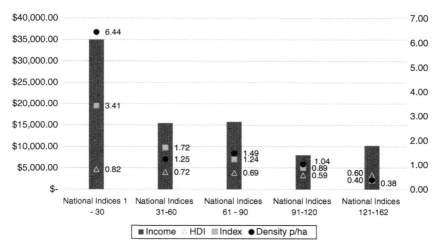

FIGURE 9.2 HDI, average income, density (persons/hectare) and index for batches of 30 countries, grouped by index.

The energy mix of each country is also a factor that affects the index. Countries with high energy consumption and/or use of fossil fuels have, in general, a higher index, but the indications are not as clear as for population density, HDI and average income.

Understanding the index and how it has affected the calculation of the environmental impact is essential before comparing the families' lifestyles and habits. The families' environmental impacts were first converted to hectares and then multiplied by the index for each country to derive the impact in global hectares. It is important to recognise how much a hectare in New Zealand can produce when compared with a hectare in Mongolia, Germany or anywhere else. When the impact is transformed into global hectares, the comparison becomes more relevant. Table 9.2 sets out the indices for all countries in the stories from highest to lowest.

Comparison of the footprints

Given that every family in the stories does similar things (eating, work/education, leisure), these conversion factors have a significant effect on the ecological footprint (EF), and mean that the German family have the highest impact, as can be seen in Table 9.3.

Table 9.3 shows that households with a more traditional subsistence lifestyle (Myanmar, India, Mozambique and Indonesia) have, as might be expected, very low environmental impacts. The difference in impact for more developed world lifestyles such as South Africa (1.64 global hectares) and Germany (10.47 global hectares) comes not so much from habits as from where people live, as Germany has a high index compared with South Africa.

TABLE 9.2 Index for each country in the stories

Countries	Index
Germany	4.78
UK	3.39
Japan	2.35
Brazil	2.13
Malaysia	2.13
India	2.00
Tonga	1.99
Finland	1.61
Indonesia	1.61
New Zealand	1.53
Myanmar	1.51
United States	1.21
Argentina	1.09
Mozambique	0.73
Cuba	0.73
South Sudan	0.63
Morocco	0.62
Canada	0.54
Australia	0.50
South Africa	0.50
Mongolia	0.27

Comparison can be made between the impact of the families in the stories and the average EF for each country as calculated by GFN (Table 9.4), noting that the GFN method is a top-down approach. Our calculations include almost all the items that make up the EF in the GFN calculations, so it is not unreasonable to make comparisons. Table 9.4 indicates that 66 per cent of our families have an impact lower than the average for their country, while the four highest EF families have impacts approximately twice the average EFs of their countries.

Breaking down the EF

The breakdown of what leads to a high impact is different for each case. For the German family, it is their driving (38 per cent of their impact), while for the English family it is their consumer goods (32.5 per cent), for the rural Brazilian family their food (41.8 per cent) and for the New Zealand family their energy consumption, at 44 per cent of their impact. The highest impact of the urban Australian family is their food, at around 83 per cent of their total EF. Despite their local and organic food, they still have a large environmental impact from food. The German family, despite being one of the few families to have a solar water heater, are particularly affected by Germany's high index. While the Mongolian family clearly have a reliance on coal, they have less than half the German impact. The great problem for the German family is driving. Their CO_2 emissions are high, and after the application

TABLE 9.3 EF of each family as calculated here in global hectares per person

Countries	Global hectares/person
Germany	10.47
New Zealand	8.84
United Kingdom	7.61
Brazil, rural	5.90
Mongolia	4.94
Malaysia	4.38
United States, rural	4.30
United States, urban	4.11
Canada	4.03
Australia	3.84
Finland	3.81
Argentina	3.43
Morocco	2.66
Brazil, urban	2.31
Japan	2.06
Tonga	1.71
South Africa	1.64
Myanmar	1.09
Cuba	0.83
India	0.33
South Sudan (UNMISS)	0.16
Mozambique	0.13
South Sudan (Juba)	0.10
Indonesia	0.02

of the index, the EF of their driving rises to 4.04 global hectares/person. They drive around 30,000 kilometres per year and travel 1,400 kilometres by train, generating around 5,000 kilograms of CO_2 per year for transport. These emissions are considerably less than those of the rural Brazilian family, who we estimate drive a total of 53,600 kilometres a year. For the calculation, the Brazilian family benefit from Brazil's lower national index and the fact that the family are made up of five people against two in Germany, meaning that the travel is shared between more people. The German family drive the second largest distance among the families investigated, with rural Brazil the highest. The impact of driving in Germany, which means using the country's infrastructure of roads, makes the environmental impact from the German family's driving alone higher than the total impact of 19 out of the 24 families investigated. This confirms that not only is driving a huge burden on the environment, but so is the infrastructure to allow people to drive everywhere, which is part of the reason why Germany has such a high index.

On the other hand, the electricity supplier chosen by the German family is exclusively reliant on renewable energy, and this leads to a very low impact of around 0.2 global hectares/person for their electricity. This remarkable domestic electricity performance is offset by the high consumption of wood and oil for heating the house which brings the impact to close to 3 global hectares/person. When

144 Fabricio Chicca, Robert Vale and Brenda Vale

TABLE 9.4 Comparison of EF as calculated here and average EF in global hectares from GFN data

Countries	EF from stories gha/person	Average national EF from GFN gha/person
Germany	10.47	4.40
New Zealand	8.84	5.10
United Kingdom	7.61	4.20
Brazil, rural	5.90	2.90
Mongolia	4.94	4.50
Malaysia	4.38	2.90
United States, rural	4.30	6.80
United States, urban	4.11	6.80
Canada	4.03	6.60
Australia	3.84	8.30
Finland	3.81	4.80
Argentina	3.43	2.80
Morocco	2.66	1.50
World average	2.65	2.65
Brazil, urban	2.31	2.90
Japan	2.06	3.80
World bio-capacity	1.72	1.72
Tonga	1.71	2.20
South Africa	1.64	2.50
Myanmar	1.09	1.50
Cuba	0.83	1.60
India	0.33	0.90
South Sudan (UNMISS)	0.16	1.35★
Mozambique	0.13	0.90
South Sudan (Juba)	0.10	1.35★
Indonesia	0.02	1.30

★Averaged global hectares from Chad, Niger, Ethiopia and Mali

it comes to total domestic energy, the New Zealand family has the highest environmental impact among all the families. This family offsets a reasonably good national energy mix with a high consumption of firewood for heating. While firewood can be considered carbon neutral, it takes land to grow trees for firewood so it will have an EF from the land occupied.

The New Zealand family scores very high in both the food and the energy categories, with the second largest impact for food and the highest score per person for domestic energy, leading to a very high overall environmental impact. Comparing the New Zealand and Australian families, the latter have a much lower impact. The Australian family benefits from a lower national index when compared with New Zealand. The low index of 0.50 for Australia (see Table 9.2) is mostly because Australia has a very low density. In addition, it is one of the driest countries, with rainfall being unpredictable (Bradtke, 2017), resulting in only 6 per cent of the total area being arable land (Millar and Fyfe, 2012). This means that a hectare of

Australian land is, on average, not very productive compared with the world average, because a large part of the country is perceived as desert. The original, sustainable, inhabitants of Australia made this perceived desert their home and for them, the land was not unproductive, but this is perhaps another story. The comparison between the Australian and New Zealand indices helps to explain why the urban Australian family scored much lower than the rural New Zealand family.

With similar indices in Morocco (0.62) and Australia (0.50), the families have very different impacts. While the Morocco family have an impact of only 2.66 global hectares, the Australian family score 44 per cent higher. Considering the HDI from both countries (0.939, ranked second, for Australia and 0.647, ranked 123rd, for Morocco [UNDP, 2016b: 22–24]), a larger difference might have been expected. However, a large share of the impact of both comes from food, with diets based on meat, and food production with similar yields. Meat consumption has a great effect on the overall environmental impact, and by a considerable margin, the New Zealand and Mongolian families eat the most meat. The Mongolian family consumes 173 kilograms of beef and mutton and 121 kilograms of milk per person a year, although this is offset by the low national index for Mongolia. The New Zealand family annually consumes 126 kilograms of meat, 156 kilograms of milk, 2.6 kilograms of butter and 32.5 kilograms of cheese per person, but when the national index is applied, even though Mongolia has an environmentally worse energy mix, with comparable meat consumption the New Zealand family has a much higher environmental impact.

Coming down the list in Table 9.4, the EF of the Malaysian family is higher than the GFN national average EF for Malaysia. Their food consumption has some animal products but is heavily dependent on fast food. Malaysia has an HDI of 0.77, a density of 0.9 persons/hectare and an index of 2.13, the latter being slightly higher than the indices of countries with the same HDI. An important aspect of Malaysia that pushes its index up is the fact that 80 per cent of the peninsular territory is covered by tropical forest (Chee and Peng, 2006). Tropical forest has the highest capacity for sequestering CO_2 (IPCC, 2007) which makes plausible the high index because one hectare of this forest will represent several hectares of global average forest.

The American families, one rural and one urban, have a similar overall EF but the breakdown is different. The highest part of the impact of the urban household comes from food, as part of the family is highly dependent on fast food. In contrast, the rural family scores very highly in transportation. The United States has a low index when compared with European countries with a similar HDI. Both American families, however, have performances better than their fellow citizens when compared with the GFN figures. The American average EF is 6.8 global hectares, while the rural family is 4.3 global hectares and the urban 4.11 global hectares. Canada is another instance where a low national index of 0.54 helps the family to score relatively low in comparison with European countries with similar HDIs. Canada has a very low density (0.03 persons/hectare), and the part of its territory in the boreal (subarctic) climate zone brings the index down. Despite a higher proportion of Finland being located in the cold zone, at 1.61 it has a higher index than Canada. The reason is much higher density (0.16 persons/hectare) and a worse energy mix.

The Japanese family had the lowest environmental impact among countries with high HDIs. Japan has an HDI of 0.903 (ranked 17th) (UNDP, 2016a: 22). The family have a low consumption of meat, dairy products and energy, and do little driving. The impact from food in Japan is partially minimised by the country's very productive agricultural land (FAO, 2017). Therefore, even with an index of 2.35, the family members have a relatively low environmental impact with an EF of 2.06 global hectares. Although Brazil is a developing country, the social condition of the country with its large social gap means the higher and at least some middle-class households have most of the privileges of developed countries. For the Brazilian urban family, diet choice, limited driving, a small house and low energy consumption produce a relatively low EF of 2.31 global hectares. In both Brazilian case studies, the highly ranked (fourth) index of 2.13 happens because almost the entire country is located in the tropical zone. A superficial analysis would not suggest a high index as Brazil has a low density of 0.24 persons/hectare and only a modest infrastructure when compared with high HDI countries. However, Brazil has very fertile land, and a vast area of the country is covered by rainforest. This suggests that land in Brazil has a very high environmental value and it would take quite a lot of global hectares to produce what can be produced from one hectare of Brazilian land.

The Tongan family have a diet partially based on meat. Tonga is an island country and its index of 1.99 comes in part from the scarcity of land, and therefore it is understandable that the EF of each household member is 1.71 global hectares. It is important to highlight that in Table 9.4, the Tongan family are the first among all the stories that could be considered sustainable as their EF falls just below the fair share cut-off of world bio-capacity divided by world population. However, as the world population grows, unless the Tongan family decrease their impact, very soon they will also become unsustainable. The South African family have a similar consumption of meat to the Tongan family, but a considerably higher consumption of dairy products, which increases the global hectares and hence the EF.

Learning from those that have little

The lower part of Table 9.4 requires special assessment. Any EF equal to or below the fair share of 1.70 global hectares can be considered sustainable. The first point of note is that the countries with low EFs do not have a very high HDI. In descending order of HDIs are Cuba (0.775) and Tonga (0.721), so both these countries have an HDI above 0.7, which is considered high. Below them are Indonesia (0.689), South Africa (0.666), India (0.642), Myanmar (0.556), and Mozambique and South Sudan (both 0.418) (UNDP, 2016b: 22–25). There are no countries with low EFs and very high HDI scores, which sends a salutary message. An HDI over 0.7 is considered to be high and over around 0.79 very high human development (GFN, 2017). Cuba and Tonga represent the sort of lifestyle that everyone could have while still living within the limits of the planet. Of course, the rising global population means that each year what each of us can have becomes less, as the available bio-capacity does not increase along with the population; in fact, it declines as more people mean that more of the land is built on.

Another important message is the effect of humanitarian tragedies. The South Sudan families are caught in the middle of a civil war, and this considerably limits their capacity to consume resources because war makes resources scarce. The calculation for South Sudan was also complicated by the lack of reliable data. The comparison of the South Sudan families shows the somewhat higher impact of the family inside the UN camp. This reflects the impact from the food distributed by the UN, which has a higher impact than locally produced food. Another tragedy is the case of Mozambique. An ongoing conflict in the central part of the country has affected the fragile local economy and especially the region where the family in the story live. The family are living on the verge of starvation. Without an end to the conflict, the family and their neighbours face a real challenge to survive.

In contrast, there are families with a low environmental impact that do not live at the limit of survival. On the contrary, they live very balanced lives within their communities, and are simply less dependent on fossil fuels and consumer goods than most of us are. Indeed, they represent ways of living that could become necessary in the future to dramatically reduce the environmental burden of humanity. Of course, the majority of the families in these situations aim to improve their quality of life, but there are many different stages of consumption that they would have to go through before becoming unsustainable. More importantly, for those directly involved with the production of their own food, the concept of limits is not an academic principle but a real fact. The perception that official bodies have towards these communities that are sustainable is uncertain, as the HDI and the majority of other metrics available to measure social development disregard low environmental impact. The fact remains that high HDI is associated with high environmental impact. These low-impact communities present a real solution to the problem of relieving environmental impact and climate change. We may all want all human communities to have a high HDI, but this may not be possible under the current way of assessing development (UNDP, 2016a). Health and education do not necessarily have to require resources and wealth, but the importance of gross national income may need to be reconsidered if HDI scores are to take account of environmental impact. HDIs can be raised without raising impact but only if those in the developed world realise that this cannot happen unless they consume a great deal less than they do at present.

Unfortunately, the problem of environmental impact is not decreasing. Human population growth creates the permanent problem of reducing the world biocapacity per person. At this stage, the only real solution that we have is to focus on low-EF communities like those in the stories and learn from them how to be sustainable. If there is a real intention to improve the quality of life of all such communities all over the world, it is important to recognise the relationship between improving social conditions and increased environmental impact. This recognition suggests two major issues. First, our metrics are wrong, as development must be assessed not only by what you have, but by the environmental impact that you produce, or will produce with increased development. The second is that there is no solution but to change developed world lifestyles to reduce our environmental impact to allow those that are in critical conditions to improve.

The Indian family offer a very interesting situation, not only because they have a very low environmental impact, but because they have a great social life. Even though India has a high index because of its high population density, the family in the story still have a very low environmental impact. They are vegetarian, do not drive a car and have no appliances. Their impact mostly comes from their food and the energy to cook it. Cuba is also a fascinating example. Because of their government and the blockage of commerce, Cubans are mostly outside of consumer society. Food production and commercialisation are controlled by the government. Appliances and consumer items are rare and very expensive for Cubans. They have to adapt their houses and, of necessity, apply the concept of cohousing. The country's economic arrangements, because of the trade blockage, made Cuba develop

TABLE 9.5 CO_2 per capita emissions for each family compared with EFs based on national indices

Country	Total CO_2 per family tonnes/year	CO_2 per capita tonnes/year	Ranking by CO_2	EF gha (country index)	Ranking by EF
United States, rural	20.87	10.44	1	4.30	7
Canada	13.52	6.76	2	4.03	9
Australia	17.29	5.76	3	3.84	10
Brazil, rural	27.24	5.45	4	5.90	4
United Kingdom	14.78	4.93	5=	7.61	3
Mongolia	14.78	4.93	5=	4.94	5
Germany	7.75	3.88	7	10.47	1
New Zealand	13.78	3.44	8	8.84	2
Argentina	12.17	3.04	9	3.34	12
United States, urban	11.22	2.81	10	4.11	8
Finland	9.33	2.33	11	3.81	11
Malaysia	13.94	2.32	12	4.38	6
South Africa	11.54	2.31	13	1.64	17
Brazil, urban	9.20	2.30	14	2.31	14
Tonga	6.29	1.57	15	1.71	16
Morocco	6.05	1.51	16	2.66	13
Cuba	6.86	1.29	17	0.83	19
Myanmar	4.89	1.22	18	1.09	18
Japan	5.43	1.09	19	2.06	15
South Sudan (UNMISS)	1.45	0.24	20	0.16	21
India	0.58	0.19	21	0.33	20
South Sudan (Juba)	1.55	0.12	22	0.01	23
Indonesia	0.39	0.08	23	0.02	24
Mozambique	0.56	0.06	24	0.13	22

TABLE 9.6 Breakdown of EFs in stories using national indices in global hectares/person

Country	Total	Food	Travel	Energy	Housing	(Appliances)	Consumer goods
Germany	10.47	0.82	4.04	2.96	0.49	(0.26)	1.89
New Zealand	8.84	3.49	0.54	3.93	0.06	(0.10)	0.72
United Kingdom	7.61	1.62	1.29	1.77	0.12	(0.34)	2.47
Brazil, rural	5.90	2.47	2.04	0.23	0.49	(0.21)	0.47
Mongolia	4.94	4.43	0.02	0.40	0.01	(0.01)	0.07
Malaysia	4.38	2.39	0.65	0.39	0.12	(0.10)	0.72
United States, rural	4.30	0.73	1.89	0.90	0.28	(0.06)	0.44
United States, urban	4.11	2.53	0.44	0.13	0.10	(0.11)	0.80
Canada	4.03	2.75	0.52	0.24	0.02	(0.06)	0.44
Australia	3.84	3.17	0.03	0.42	0.05	(0.02)	0.15
Finland	3.81	0.85	0.33	0.58	0.06	(0.24)	1.75
Argentina	3.43	1.75	0.31	0.38	0.08	(0.11)	0.80
Morocco	2.66	2.27	0.00	0.12	0.03	(0.03)	0.21
Brazil, urban	2.31	0.63	0.14	0.71	0.08	(0.09)	0.65
Japan	2.06	0.11	0.13	0.64	0.11	(0.13)	0.94
Tonga	1.71	0.79	0.05	0.20	0.08	(0.07)	0.51
South Africa	1.64	1.23	0.04	0.18	0.02	(0.02)	0.15
Myanmar	1.09	0.48	0.56	0.02	0.02	(0.00)	0.00
Cuba	0.83	0.71	0.01	0.08	0.02	(0.00)	0.00
India	0.33	0.12	0.00	0.20	0.01	(0.00)	0.00
South Sudan (UNMISS)	0.16	0.13	0.00	0.04	0.00	(0.00)	0.00
Mozambique	0.13	0.11	0.00	0.02	0.00	(0.00)	0.00
South Sudan (Juba)	0.10	0.07	0.01	0.03	0.00	(0.00)	0.00
Indonesia	0.02	0.01	0.00	0.01	0.00	(0.00)	0.00

a sustainable way of living. Although the country is dense (1 persons/hectare), has a high HDI (0.78) and a good infrastructure inherited from the Soviets, that currently lacks maintenance, and has an environmentally poor energy mix, it still has a national average EF on the limit of currently being sustainable. The cities are buzzing, people have their house doors open, fix their appliances when they break and have a very active social life.

The family from Myanmar also have a very active social life with no shortage of food. The two girls go to school and the family have benefited from the influx of tourists. Even without tourism, the family believe that they would still have what they would call a good life. However, our champions in sustainability are the Indonesian family. Their cultural values and traditions, which allow them no modern technology or fossil fuels, and the fact that they grow their own food using traditional methods, make them the most sustainable family among the 24 stories.

Emissions from daily habits

Another way to assess the families' habits is to consider their CO_2 emissions. The calculation for CO_2 emissions is more direct and is the way in which EFs for

TABLE 9.7 Breakdown of CO_2 emissions per capita

	CO_2 per capita tonnes/year	Food	Travel	Energy	Housing	(Appliances)	Consumer goods
United States, rural	10.44	0.47	4.69	3.58	1.03	(0.08)	0.58
Canada	6.76	0.94	2.87	1.41	0.09	(0.18)	1.27
Australia	5.76	0.73	0.16	3.74	0.46	(0.08)	0.59
Brazil, rural	5.45	0.24	2.87	0.46	0.91	(0.12)	0.84
United Kingdom	4.93	0.12	1.14	2.68	0.16	(0.10)	0.73
Mongolia	4.93	3.93	0.19	0.88	0.19	(0.03)	0.23
Germany	3.88	0.24	2.54	0.10	0.34	(0.08)	0.58
New Zealand	3.44	0.64	1.06	1.19	0.17	(0.05)	0.34
Argentina	3.04	0.32	0.84	0.94	0.29	(0.08)	0.57
United States, urban	2.81	0.26	1.08	0.52	0.38	(0.07)	0.50
Finland	2.33	0.07	0.61	0.60	0.12	(0.11)	0.82
Malaysia	2.32	0.07	0.91	0.88	0.07	(0.05)	0.35
South Africa	2.31	0.05	0.25	1.65	0.18	(0.02)	0.29
Brazil, urban	2.30	0.05	0.20	1.37	0.16	(0.06)	0.46
Tonga	1.57	0.49	0.08	0.39	0.16	(0.05)	0.40
Morocco	1.51	0.16	0.00	0.82	0.20	(0.04)	0.29
Cuba	1.29	0.59	0.04	0.48	0.11	(0.01)	0.05
Myanmar	1.22	0.04	1.11	0.01	0.04	(0.00)	0.00
Japan	1.09	0.04	0.16	0.39	0.20	(0.03)	0.25
South Sudan (UNMISS)	0.24	0.12	0.00	0.12	0.00	(0.00)	0.00
India	0.19	0.06	0.00	0.12	0.01	(0.00)	0.00
South Sudan (Juba)	0.12	0.06	0.03	0.03	0.00	(0.00)	0.00
Indonesia	0.08	0.01	0.00	0.07	0.00	(0.00)	0.00
Mozambique	0.06	0.01	0.00	0.04	0.01	(0.00)	0.00

travel and consumer goods are assessed. The combination of global hectares and CO_2 emissions provides additional data with which to compare habits rather than impact. Table 9.5 shows the CO_2 emissions for each story and gives rankings for both EF based on national indices and CO_2 emissions.

Breakdown of EFs

To understand the big impacts in terms of EF and CO_2 emissions, Tables 9.6 and 9.7 break down the EF values from Table 9.4 into their component parts. The stories from the various countries are ranked from highest to lowest impact.

No particular patterns emerge from Tables 9.6 and 9.7 apart from the fact that in the developed world, scores tend to be higher for all components than in the less developed or developing world. The following chapters look at each component of this breakdown in more detail.

References

Bradtke B (2017) *Australian desert animals*, available at http://www.outback-australia-travel-secrets.com/australian-desert-animals.html, accessed 4 September 2017.

Food and Agriculture Organization (FAO) (2017) *FAOSTAT*, available at http://www.fao.org/faostat/en/#home, accessed 21 August 2017.

Global Footprint Network (2017) *Making the Sustainable Development Goals consistent with sustainability*, available at http://www.footprintnetwork.org/2017/09/01/making-sustainable-development-goals-consistent-sustainability/, accessed 6 September 2017.

IPCC (2007) *Climate Change 2007, The Fourth Assessment Report, The Physical Science Basis*, Cambridge: Cambridge University Press.

Millar R and Fyfe M (2012) *The Fertile Fringe*, available at http://www.theage.com.au/victoria/the-fertile-fringe-20120525-1zasy.html, accessed 16 December 2017.

United Nations Development Programme (UNDP) (2016a) *Human Development Index (HDI)*, available at http://hdr.undp.org/en/content/human-development-index-hdi, accessed 4 September 2017.

United Nations Development Programme (UNDP) (2016b) *Human Development Report: Overview 2016*, available at http://hdr.undp.org/sites/default/files/HDR2016_EN_Overview_Web.pdf, accessed 4 September 2017.

World Bank (2017) *Land Area (sq. km.)*, available at https://data.worldbank.org/indicator/AG.LND.TOTL.K2, accessed 16 December 2017.

10
FOOD

Fabricio Chicca, Robert Vale and Brenda Vale

Introduction

The analysis of all the stories shows the importance that food has for the families, confirming that no matter the social condition of the families, food plays an important role in their routine. The calculation of each family's environmental impact has also confirmed the importance of food in the composition of people's total impact. All over the world, as humans have become predominantly urban, people are reducing their knowledge about the land needed to produce their food.

Research has suggested the significance of the impact caused by food production but the size of the impact is still a matter of debate. Goodland and Anhang (2009) claim that livestock contribute nearly 51 per cent of the world's greenhouse gas (GHG) emissions, and the FAO estimates that livestock's contribution is 18 per cent, while Garnett (2011) argues that there are no studies that quantify GHG emissions from the entire food chain. Different methodologies will come up with different ways to measure the impact from food production and consumption, so the proportion of the impact that originates from food will always differ from study to study. A full life cycle analysis (LCA) from farm to plate would be the best way to produce a definitive assessment of the environmental impact from food, but some major factors might not be fully included in the LCA calculation.

The matter of locality

To calculate the environmental impact from food consumption of the families in this book, we faced a major challenge: to identify the origin of the food. The food chain today is incredibly complex. Food companies are truly international businesses and food is being shipped around the world as never before. The distance that food travels takes space in the news media (Kemp et al., 2010: 504) and there is growing public interest in the impact from local or non-local food.

Coley et al. (2009) have suggested that local food is a simple, appealing concept and they have identified the relationship that consumers are making between local food and sustainability. Awareness about the distance that food travels from producer to plate has been called food miles (Kemp et al., 2010). More importantly, the public (and occasionally some authorities) have considered the concept of food miles an instrument to measure sustainability. The logistics to bring food from an overseas production location can be roughly imagined as shown in Figure 10.1.

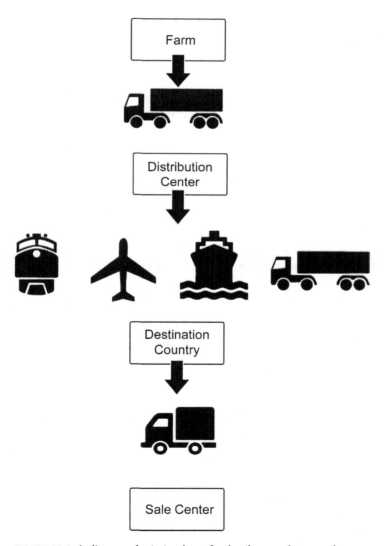

FIGURE 10.1 A diagram depicting how food miles may be caused.

Figure 10.1 does not consider all parts of the process of food production and distribution. For a consumer, it is easy to imagine that transportation is the largest environmental impact caused by food from far away; each transportation mode added to the food process produces more emissions, and therefore more impact. The rationale is right, but the assessment of sustainability is not. The transport assumptions in Figure 10.1 provide a highly simplified snapshot of the entire process. Edwards-Jones (2010) suggests that 50 kilometres is perceived as an acceptable distance for most consumers. However, as we shall see in the chapter on travel, all food miles are not equal. Transporting something 7,000 kilometres by ship could produce the same emissions as taking it 100 kilometres by road in a van.

Transport is not the only component of food emissions, and another environmental impact that is often not considered is the demand for water to produce food. Vanham et al. (2013) have stated that 84 per cent of the water consumption in the EU is directly related to food production.

The vast majority of the families included in this research had no concerns about the environmental impact caused by their eating habits. The rural American family, who have been affected by Hurricane Katrina, and the Australian family show some concerns about the impact caused by their eating habits. Indeed, the Australian family appear to have sustainability at the top of their agenda, including their food. They only buy certified organic food, prioritise local producers, and include vegetables in most of their meals. However, for most of the stories in this book, food and sustainability have no strong correlation as food is not perceived as relevant to the environmental impact.

The real environmental impact of food: local matters but does not always mean sustainable

In our research, only the rural American family and the Australian family were openly concerned about the location where the food was produced. We are sure that the families from Mozambique, Indonesia, Myanmar, South Sudan and India consume the vast majority of their products from local sources. For another group of families, which encompasses Argentina, rural Brazil, Cuba, Mongolia and South Africa, there is an indication that the food they consumed is mostly locally produced, whereas the other families are probably consumers of more globalised food.

Kemp et al., in their research among consumers in the United Kingdom in 2010, identified that price, not locality, is still the main driver of food choices. For UK consumers who bought local food, the greatest motivations were to support local producers, or simply because local is tastier. Those whose responses were related to the environment said that they wanted to reduce the food miles and to stop transportation across the world (Kemp et al., 2010). According to Coley et al. (2009), if a shopper drives a round-trip distance of more than 6.7 km alone in a car in order to purchase their organic vegetables, it is likely that their overall emissions will be greater than for the conventional food system.

The environmental impact from food is not measured only by the distance that it has covered to reach the plates. Transportation might be a relevant part of the environmental impact from food, but normally it is not the major problem. Moudry et al. (2013) show that the production of a kilogram of potatoes is estimated to produce 0.145 kilograms of CO_2 in the Czech Republic. Edwards-Jones (2010) gives the emission of 0.005 kilograms of CO_2 to move a tonne of goods one kilometre by ship, 0.037 kilograms of CO_2 per tonne-kilometre by train and 1.076 kilograms CO_2 for a tonne-kilometre transported by road in a van. If a kilogram of potatoes produced near Prague has been shipped by train to Hamburg (approx. 640 kilometres), then shipped again by sea from Hamburg to London (approx. 750 kilometres) and finally distributed by a van (20 kilometres), it will have an additional 0.049 kilograms of CO_2. Therefore, a kilogram of Czech potatoes consumed by a Londoner will have an approximate emission of 0.194 kilograms of CO_2.

On the other hand, a kilogram of organic potatoes produced near Brno and transported by van to Prague (205 kilometres) would have an additional 0.22 kilograms of CO_2. As organic potatoes have an emission of 0.126 kilograms CO_2 (Maudry et al., 2013) the total emissions for a kilogram of potatoes for a citizen in Prague eating "local" organic potatoes from Brno will be 0.346 kilograms of CO_2. In this case, for a Czech citizen to eat fairly locally, at least from their own country, organic potatoes would cause more emissions than for someone overseas eating conventional potatoes grown near Prague. The distance is important, but the way that we move the goods around is actually much more important. The miles are less important than the mode of transport.

Examining the potatoes eaten in London, even moving the potatoes more than 1,400 kilometres, the largest part of the environmental impact has happened inside the farm gate. A similar study conducted by Brodt et al. (2013) compared the environmental impacts of tomato-derived products produced locally in Michigan or shipped from California. Emissions from transport by rail are offset by much higher productivity in California, but this is only possible with heavy irrigation, requiring six times more irrigation than Michigan.

Another way to consider the importance of locality, or food miles, is to consider the diagram in Figure 10.2. It indicates the estimated amount of CO_2 emissions from each transportation mode and in what part of the food's journey it might be used. The emissions refer to kilograms of CO_2 per tonne-kilometre.

People worry much more about food miles by air than they do about food miles by van, although both have similar emissions. Miles matter very little if food travels by ship.

Energy in, energy out and the human weakest spot

Energy is essential to produce food, but how much energy is necessary to produce the food, from the farm to the plate, compared with the amount of energy that the food generates? As food can be measured by the energy it provides us when we eat it, usually measured in kcal (kilocalories), it is easy to produce an indicator

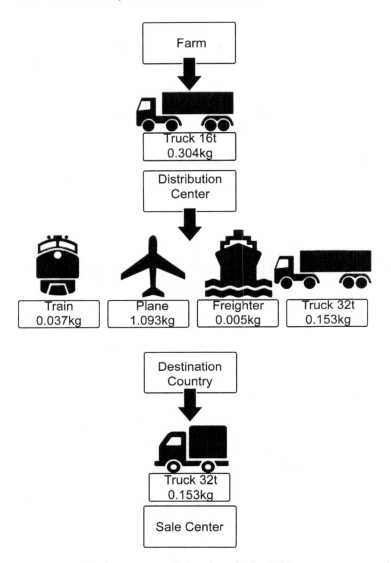

FIGURE 10.2 Food transport emissions by vehicles in kilograms per tonne-kilometre.

comparing the amount of energy to produce the food and the energy provided when we eat it. Treloar and Fay (1998) suggested that 42 gigajoules of energy (nearly 12,000 kilowatt hours) per adult per year is necessary to provide food in Australia. This figure includes not only the food itself, but every single part of the process from farm to table. The daily food intake for an adult male should be 10,500 kilojoules (NHS, 2016b), which is 3.8 gigajoules or 1,066 kilowatt hours a year, so the energy in is more than ten times the energy out.

Another way to relate food and energy is the assessment of productivity to grow food per hectare measured by energy outputs from the food. Argentina has an average yield of 2,667 kilograms/hectare/year for wheat and 29,441 kilograms/hectare/year for potatoes. Considering that a kilogram of wheat will provide approximately 3,320 kilocalories (USDA, 2017), a hectare in Argentina will provide about 37 gigajoules of food energy. On the other hand, potatoes will provide 95 gigajoules. Therefore, the comparison between wheat and potatoes for Argentina suggests that producing potatoes in Argentina will generate more than twice as much energy as wheat would from the same area of land.

A more interesting analysis can be made when comparing plant products with meat and milk. The amount of energy produced from meat by a hectare in comparison with potatoes, wheat, tomatoes and soy is much lower. Table 10.1 shows how many times more energy plant crops are capable of producing when compared with red meat using the same amount of land in the same country.

Our assumptions have been heavily based on GFN and FAO reports, which in both cases do not include the amount of direct energy input to produce the agricultural items. Similarly, Reijnders and Soret (2003) have suggested that in industrial countries, meat demands between 6 and 17 times more land to produce the same amount of energy than soy-based items.

TABLE 10.1 The amount of energy that could be produced from different food types compared with one unit of energy from meat

	Meat	Potatoes	Wheat	Tomatoes	Soybeans
Japan	1	15.3	8.5	7.0	1.6
Germany	1	30.3	704.6	38.2	2.4
United States	1	40.3	10.8	17.7	5.2
Brazil	1	46.3	15.8	25.9	9.1
United Kingdom	1	10.1	287.8	164.4	NA
Cuba	1	44.6	NA	5.1	NA
Myanmar	1	32.9	17.8	NA	4.3
South Africa	1	100.6	45.3	52.6	10.3
Tonga	1	0.0	NA	23.1	NA
Finland	1	82.7	53.2	266.7	NA
New Zealand	1	160.4	122.6	90.2	NA
Argentina	1	112.9	44.1	37.5	20.3
Mongolia	1	49.4	29.3	NA	NA
Canada	1	141.0	57.0	73.3	22.1
Morocco	1	155.2	37.3	92.5	9.3
Australia	1	207.9	45.2	70.5	21.6
Indonesia	1	95.6	NA	19.6	16.0
Malaysia	1	NA	NA	151.5	NA
India	1	217.1	128.5	47.0	17.5
Mozambique	1	110.9	44.5	47.6	NA

The table shows how much energy could be produced by comparing different food types with one unit of energy from meat. The yield data were extracted from the FAO (2006) while the energy data come from the USDA (USDA, 2017).

Table 10.2 compares milk with various plant crops. Table 10.2 does not include the amount of energy necessary to produce milk. To achieve a high-yield milk production in some areas relies on mechanised production and on grain supplements to feed animals. Plant crops may require more energy than extensively raised grass-fed cattle, but in most cases, the situation will favour the plant crop as the data above show. In all cases, plant crops would have a better energy output than red meat and milk, except in the case of soybeans in Germany and Japan compared against milk.

Neither the FAO nor the USDA provide the amount of energy necessary to produce food. The FAO provides data regarding GHG emissions, but not the breakdown to identify the crops or livestock (with details given only for some products which are high contributors to emissions, such as rice). However, there are overwhelming indications that red meat and the dairy industry are inefficient when comparing the amount of energy in with energy out, particularly when compared with other agricultural products that are not related to livestock. Webber (2011) has suggested that dairy, meat and eggs have the worst ratio of energy input to output compared with grains, vegetables, fruit and tree nuts. A more detailed set of data confirming the energy efficiency from non-animal crops comes from Pimentel (2009). Pimentel has compared different crops, considering energy in to energy out, including labour, machinery, fossil fuel, fertiliser, irrigation, electricity and transport as energy

TABLE 10.2 Milk compared with various plant crops

	Milk	Potatoes	Wheat	Tomatoes	Soybeans
Germany	1	4.84	112.75	6.11	0.39
Japan	1	4.36	2.40	1.99	0.47
Tonga	1	NA	NA	2.78	NA
United Kingdom	1	2.28	65.27	37.28	NA
New Zealand	1	22.89	17.49	12.87	NA
India	1	12.67	7.50	2.75	1.02
Finland	1	24.86	15.99	80.16	NA
Myanmar	1	14.47	7.85	NA	1.90
United States	1	52.71	14.16	23.22	6.82
Cuba	1	51.15	NA	5.83	NA
Brazil	1	95.07	32.39	53.17	18.61
Canada	1	128.19	51.77	66.62	20.05
Indonesia	1	109.15	NA	22.42	18.30
Morocco	1	200.78	48.28	119.67	11.98
Argentina	1	285.36	111.57	94.84	51.36
Malaysia	1	NA	NA	265.95	NA
South Africa	1	839.93	378.13	439.51	85.95
Australia	1	1496.61	325.51	507.48	155.33
Mongolia	1	723.22	428.39	NA	NA
Mozambique	1	7270.29	2918.39	3120.97	NA

The table shows how much energy could be produced by comparing different food types with one unit of energy from milk. The yield data were extracted from the FAO (2006) while the energy data come from the USDA (USDA, 2017).

TABLE 10.3 The energy efficiency (energy in/energy out) of crops, adapted from Pimentel (2009)

	Kcal input/kcal output	Country
Potatoes	1/2.76	United States
Wheat	1/3.31	Kenya
Tomatoes	1/0.78	United States
Soybeans	1/3.71	United States

inputs against the data from the calories produced for different crops as energy out. Table 10.3 shows the relation between energy in and energy out for different crops.

Table 10.3 shows that plant crops can often require less energy input than they yield as food energy. Pimentel and Pimentel (2003) found that this input/output ratio is overwhelmingly unfavourable for meat. Beef cattle require 40 times more energy than the energy generated. Milk has a better ratio but is still poor compared with other food items. Milk requires 13 times more energy to produce than the energy provided to those who drink it. Reijnders and Soret state that meat demands between 6 and 20 times more fossil fuel to produce the same amount of energy than soybeans. MacKay (2009) also suggests that meat has very low efficiency (4.3 per cent), which means roughly 25 times less energy is provided than is needed to produce it. However, there is a better result for milk (45 per cent), meaning that each glass of milk provides almost half of the energy necessary to produce it. In spite of differences between production methods (extensive and intensive), climate, animal species and so on, meat and dairy products will have a negative output of energy compared with non-animal-based food items.

Red meat and dairy are the largest contributors among the livestock but the remaining farm animals contribute to GHG emissions. In terms of fossil fuel energy requirements, de Vries and de Boer (2010) have suggested that pork and chicken are the most efficient, using from 18 to 45 megajoules/kilogram for pork and 15 to 29 megajoules/kilogram for chicken, whereas beef requires from 34 to 52 megajoules/kilogram. To put these figures into perspective, a kilogram of meat will contain between 4 and 6 megajoules (averaged from data on a variety of raw cuts of beef, pork and chicken from NHS, 2016a), so the energy input ranges from three to nine times the energy you get from eating the meat. De Vries and de Boer (2010) found that to produce a kilogram of protein, milk requires from 37 to 144 megajoules/kilogram and eggs require 87 to 107 megajoules/kilogram while beef requires from 177 to 273 megajoules/kilogram of protein. Chicken and pork have figures in between milk and beef. A gram of protein from any source contains 4 kilocalories (UCLA, 2005: 3) so a kilogram of protein will be 4,000 kilocalories or 16.75 megajoules. What these energy figures show, since they are based on fossil fuels needed to produce these foods, is that in all cases more energy is needed to produce the foods than we obtain from eating them, so when we eat, we are in a very real sense eating fossil fuels. Given that these fuels are finite, this would appear to be a concern.

Meat and dairy production has an additional impact, which is perhaps the largest issue from meat and dairy products. Inside the farm gate, the prevailing GHGs are nitrous oxide (N_2O) from soil preparation and also from manure and urine from livestock, and methane (CH_4) from ruminant digestion. According to the IPCC (2007), a kilogram of CH_4 may have over 70 times more global warming potential (GWP) in a period of 20 years than a kilogram of CO_2. Agriculture contributes about 47 per cent and 58 per cent of total anthropogenic emissions of CH_4 and N_2O, with the largest contribution coming from livestock-related emissions. Garnett (2011) also points out the impact from the emissions induced by land change to produce either meat or dairy, or grains to feed livestock (which are not included in the calculation of emissions from the FAO). It is estimated that there is up to a 17 per cent addition in emissions from the change of land use either to grow cattle or to produce grains to feed livestock.

The production of meat and dairy products results in CO_2 equivalent ($CO_{2\text{-e}}$) emissions as a result of N_2O and CH_4. Providing 1 kilogram of pork produces something between 3.9 and 10 kilograms $CO_{2\text{-e}}$, while for a kilogram of chicken, the figure is between 3.7 and 6.9 kilograms $CO_{2\text{-e}}$. In contrast, the figure for red meat goes from 14 to 32 kilograms $CO_{2\text{-e}}$ per kilogram. This difference between the different livestock can be explained. The emission of CH_4 from monogastric animals such as pigs and chickens comes only from their manure, and typically represents only 25 per cent of the total, while the other 75 per cent of emissions comes from enteric fermentation in the rumens of ruminants such as cows and sheep (Goldewijk, Olivier & Peters, 2005). De Vries and de Boer (2010) have concluded that 1 kilogram of red meat has the largest contribution to environmental impact followed by pork, chicken, eggs and milk. While milk may have a considerably lower impact, dairy products may cause a greater impact. For instance, cheese may have upto eight times more impact than milk itself, but it is still lower than the environmental impact from meat, which is around 11 times higher than milk (Vale & Vale, 2009: 40).

To return to the Czech potatoes, Moudry et al.'s (2013) investigation suggested GHG emissions of 0.145 kilograms/CO_2 for each kilogram of potatoes, while the figure for emissions for meat from the Czech Republic from FAO (2006) is a staggering 27.4 kilograms of CO_2 per kilogram.

Results

The overall picture

Although it is important to stress the differences in climate, location, ecosystems and overall environmental impact, it is possible to obtain a picture of the ecological impact from the families assessed. Considering both ecological footprint (EF) and carbon footprint, the largest environmental impact is related to the consumption of animal origin products. In order to keep the analysis consistent, despite its limitation, we have chosen to consider that the food was locally produced (within the country boundaries). For the majority of the countries, the data were collected from the FAO. Another important source of data has come from the Global Footprint Network (GFN). Through the biocapacity figures provided by the GFN, we were able to calculate the index for each country, which allows us to convert from hectares to global hectares (see further details in the calculation chapter). In most cases, a more complete analysis can be made by looking at the EF and using the CO_2 emissions as supportive data. The relation between eating and CO_2 emissions and EF is complex, but we can say that livestock products are the largest environmental impact contributors. Factors such as yield, quality of the soil, availability of water, climate, demand of energy and energy mix, size of the country, population, intensive or extensive production, amount of cropland dedicated to feed animals, quantity of pasture available, national environmental policies, quality of labour and the need to amend the soil all affect the final global hectares. The best way to have a clearer picture of the environmental impact of food is to take a look at the numbers for EF in global hectares and CO_2 emissions in Figures 10.3 and 10.4 respectively.

In general, there is a strong relationship between consumption of animal products, EF in global hectares and CO_2 emissions. In some cases, good productivity and moderate consumption of meat may generate a lower impact, as can be seen in the case of Japan and Germany. On the other hand, the families from New Zealand and Australia have a huge consumption of livestock products; in fact, the New Zealand family tops the list in terms of total consumption of red meat. Kiwis benefit from very productive land for cattle, with the herds grass-fed, a very efficient agricultural industry and an energy source mix cleaner than Australia (and much cleaner than Mongolia). On the other hand, the Mongolian family represent the largest total consumption of meat—they consume on average 173 kilograms of meat per person (red

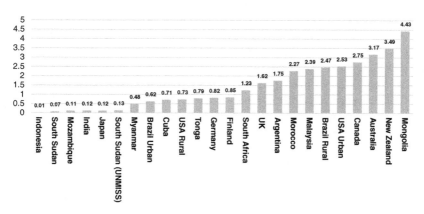

FIGURE 10.3 The environmental impact for food in global hectares per person for a year.

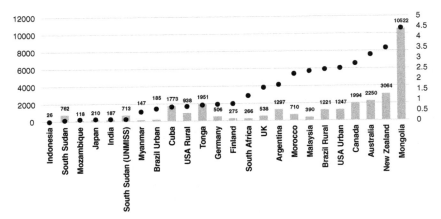

FIGURE 10.4 The environmental impact for food in CO_2 and ecological footprint. The bars express the results for CO_2 in kilograms per person (left hand vertical scale), while the dots show the impact in global hectares per person (right hand vertical scale).

meat and mutton) annually. More importantly, they have by far the largest CO_2 emissions. Despite the quantity of meat and milk intake for the family, their EF (global hectares) is proportionally only marginally higher than the other families, while the CO_2 emissions are much higher. This is likely because cattle in Mongolia are raised by nomadic groups, on natural pasture, using cattle species that are able to stay in the pasture through the four seasons, regardless of the harsh winter. The combination of severe weather and very low rainfall (257 millimetres a year) makes the conditions for crops challenging and makes it desirable to farm cattle, despite the low commercial productivity (Mongolian Statistical Information Service, 2014). A third of the agricultural land in Mongolia is abandoned, providing a very high bio-capacity per capita for the country, even with land inadequate for agriculture. The result of the Mongolian case was proportionally low global hectares and very high CO_{2e} emissions. In this case, the amount of CH_4 and N_2O determine part of the impact from the Mongolian family. It is a good example, as the combination of two sets of data is relevant to the overall picture.

Another factor that affects the impact from food is the amount of eating out, particularly fast food. Normally, fast food represents a higher environmental impact (see the calculation method). According to Vale and Vale (2009), from the Cardiff study, 90.1 per cent of the food consumed was home-made food, representing only 67.6 per cent of the environmental impact, and the other 9.9 per cent of the weight, which was food eaten out, represented 32.4 per cent of the global hectares for food (Vale & Vale, 2009: 42). The same proportion as Cardiff was considered in our study. The same additional proportion was added to the CO_2 emissions calculations. Fast food has, in general, a larger environmental impact as it is mostly very processed food, and therefore requires a greater amount of energy to be prepared, transported, cooked and served.

In some cases, as for the United States, the eating-out habits of the urban family are clearly demonstrated in the impact on global hectares. Additionally, the rural American family have a considerably higher consumption of meat. Despite eating a smaller amount of red meat than the rural family, the urban Americans have a larger intake of processed food and diary—they have the highest amount of milk intake, along with New Zealand.

The German family have a high consumption of meat, but despite this, the impact from the German family is slightly below the German average for food, but still over the country's bio-capacity.

Canada, Malaysia and the American urban family had their impact considerably affected by the amount of fast food they had. For those families, fast food is not an exception in their routine; on the contrary. The Canadian and the urban American families, in particular, rely on fast food as an effective part of their diet. There are some interesting cases to be noted. The urban Brazilian family, from the middle class, which has access to a wide variety of diets, has intentionally chosen a traditional diet, heavily based on beans and rice and having meat as a complement. A very balanced diet, with a low intake of animal products, has brought their impact to a relatively low number. Table 10.5 below indicates that Germany, Japan and rural United States have a consumption of meat relatively high compared with their global hectares and CO_2 emissions. The research suggests that, in each case, the local industry must be highly productive, minimising the impact on land consumption, but potentially causing impacts in different areas, such as water and land change.

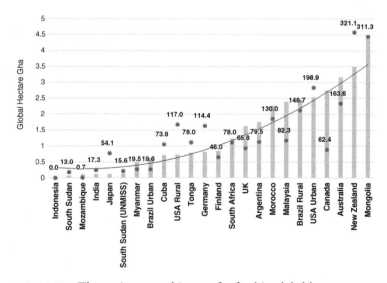

FIGURE 10.5 The environmental impact for food in global hectares per person and the annual consumption of animal products (including all kinds of meats, dairy products and eggs [expressed in kg per person]). The bars show the impact in global hectares and the dots with units show the amount of animal products in kilograms.

The bright side

Looking at the countries with lower environmental impacts, and excluding the humanitarian catastrophes, there are several important things we can learn from those countries. In India and Indonesia, the families are engaged with food production. Those families appreciate the principle of limits by directly participating in the production of their own food. The concept of limits is perhaps the most important environmental principle disregarded by the modern urban middle class and wealthy families who source food from all over the world. In most cases, from this research and several other studies, food represents the largest environmental impact per capita. Some families have chosen to reduce their intake of animal products. Despite a more modest life, the families from India and Indonesia are mostly able to fulfil most of their nutrition needs with a very low environmental impact. Despite the fact that some lifestyles in this research would be considered undesirable by city dwellers, families in developed countries, in general, have unsustainable eating habits. The environmental impact from food is not only linked with country or social level, it is related to eating habits and personal choices. In most cases, it is a matter of choice, and the choices are mostly not driven by environmental issues. Among poor families, the absence of meat, which has the highest environmental impact, is not necessarily a choice—they have a lack of options for food. Among wealthier families, meat is often the main dish, regardless of the environmental impact. Frequently, wealthier families choose local organic meats, overlooking the real impact of those items.

Limitations

Two possible limitations need to be mentioned, water footprint and waste disposal. For both water and waste disposal, we could not find a set of data that could be used across the borders without major adjustments. Vourch et al. (2008) have indicated that the dairy industry is the greatest water polluter among the food industries, generating from 0.2 to 10 litres of effluent per litre of milk. Gerbens-Leenes et al. (2013) pointed out that the water footprint for pork is four times larger than for grains. The overwhelming amount of literature on water consumption and waste disposal supports the overall calculation from our research that animal products cause the largest environmental impact.

Conclusions

Livestock production is the largest user of land, directly through grazing or indirectly through production of grains to feed to animals (Alexandratos & Bruinsma, 2012). The world population in 2030 is estimated to reach 8.5 billion people. For all nations, the estimation for consumption of meat is to increase 8.3 per cent to 2030 (43.9 kilograms of meat per year). By 2030, highly developed nations are expected to be consuming 89 kilograms of meat per person per year, while developing nations will be expecting to consume 37 kilograms of meat per person per year (FAO, 2006). Therefore, in order to keep the current environmental impact, we will have to improve productivity. To stay within the limits relies on developing countries consuming considerably less meat and milk than the rich. As result, the increased consumption of meat and dairy products in the wealthy countries becomes not only an environmental problem, but also an issue of global equity. Currently, the families in several countries are already beyond the developed nations' meat and milk thresholds for 2030. For meat, this includes the families from New Zealand, Mongolia, urban United States, Australia, rural Brazil, Morocco, rural United States and, potentially, Canada, depending on the fast food. For milk, families from New Zealand, the United Kingdom, urban United States, and Mongolia have already passed or are about to cross the expectations for rich nations in 2030.

There are not many options, and the conclusion is simple. The consumption of products from livestock must be dramatically reduced and a more vegetable-based diet must be re-introduced in order to accommodate the population growth expected for the next 20 years.

References

Alexandratos N and Bruinsma J (2012) *World Agriculture Towards 2030/2050*, Rome: FAO.
Brodt S, Kramer K J, Kendall A and Feenstra G (2013) Comparing environmental impacts of regional and national-scale food supply chains: A case study of processed tomatoes, *Food Policy*, 42, pp. 106–114.
Coley D, Howard M and Winter M (2009) Local food, food miles and carbon emissions: A comparison of farm shop and mass distribution approaches, *Food Policy*, 34(2), pp. 150–155.
Edwards-Jones G (2010) Does eating local reduce the environmental impact of food production and enhance consumer health?, *Proceedings of the Nutrition Society*, 69(4), pp. 582–591.
FAO (2006) *Livestock's Long Shadow*, Rome: FAO.
FAO (2017) *FAO data*, available at http://www.fao.org/faostat/en/, accessed 12 August 2017.
FAO and Government of Mongolia (2012) *Country Programming Framework*, Rome: FAO.
Garnett T (2009) Livestock-related greenhouse gas emissions: Impact and options for policy makers, *Environmental Science & Policy*, 12(4), pp. 491–503.
Gerbens-Leenes P, Mekonnen M and Hoekstra A (2013) The water footprint of poultry, pork and beef: A comparative study in different countries and production systems, *Water Resources and Industry*, 1, pp. 25–36.
Goldewijk K K, Olivier J and Peters J (2005) *Greenhouse Gas Emissions in the Netherlands*, Amsterdam: National Institute for Public Health and the Environment (RIVM).
Goodland R and Anhang J (2009) *Livestock and Climate Change. What If the Key Actors in Climate Change Are Cows, Pigs, and Chickens?* Washington, DC: WorldWatch Institute.
IPCC (2007) *Climate Change 2007: Working Group I: The Physical Science Basis*, 11 August 2007, available at https://www.ipcc.ch/publications_and_data/ar4/wg1/en/ch2s2-10-2.html/, accessed 12 August 2017.
Kemp K, Insch A, Holdsworth D K and Knight J G (2010) Food miles: Do UK consumers actually care? *Food Policy*, 35(6), pp. 504–513.
MacKay D (2009) *Sustainable Energy – Without the Hot Air*, Cambridge: UTI Cambridge.
Mongolian Statistical Information Service (2014) *Annual average temperature and precipitation*, 9 September 2014, available at http://www.1212.mn/statHtml/statHtml.do, accessed 27 August 2017.

Moudry J M, Jelínková Z, Jarešová M, Plch R, Moudrý J and Konvalina P (2013) Assessing greenhouse gas emissions from potato production and processing in the Czech Republic, *Outlook on Agriculture*, 42(3), pp. 179–183.

NHS (2016a) What should my daily intake of calories be? *NHS Choices*, available at http://www.nhs.uk/chq/Pages/1126.aspx?CategoryID=51&SubCategoryID=165, accessed 30 September 2017.

NHS (2016b) Calorie checker, *NHS Choices*, available at http://www.nhs.uk/Livewell/weight-loss-guide/Pages/calorie-counting.aspx, accessed 7 September 2017.

Pimentel D (2009) Energy inputs in food crop production in developing and developed nations, *Energies*, 2(1), pp. 1–24.

Pimentel D and Pimentel M (2003) Sustainability of meat-based and plant-based diets and the environment, *The American Journal of Clinical Nutrition*, 78(3), pp. 660–663.

Reijnders L and Soret S (2003) Quantification of the environmental impact of different dietary protein choices, *The American Journal of Clinical Nutrition*, 30(1), pp. 6645–6685.

Treloar G and Fay R (1998) The embodied energy of living, *Environment Design Guide GEN 20*, pp. 1–10, Sydney: Royal Australian Institute of Architects.

UCLA (2005) *Calories Count*, Los Angeles: The Regents of the University of California.

United Nations (2017) *Sustainable Development Goals Home*, available at http://www.un.org/sustainabledevelopment/blog/2015/07/un-projects-world-population-to-reach-8-5-billion-by-2030-driven-by-growth-in-developing-countries/, accessed 16 August 2017.

USDA (2017) *USDA food composition databases, August 2017*, available at https://ndb.nalusda.gov/ndb/serch/list, accessed 9 August 2017.

Vale R and Vale B (2009) *Time to Eat the Dog? The Real Guide to Sustainable Living*, London: Thames and Hudson.

Vanham D, Mekonnen M M and Hoekstra A Y (2013) The water footprint of the EU for different diets, *Ecological Indicators*, 32, pp. 1–8.

Vourch M, Balannec B, Chaufer B and Dorange G (2008) Treatment of dairy industry wastewater by reverse osmosis for water reuse, *Desalination*, 219(1–3), pp. 190–202.

de Vries M and de Boer I (2010) Comparing environmental impacts for livestock products: A review of life cycle assessments, *Livestock Science*, 128(1), pp. 1–11.

Webber M E (2011) More efficient foods; less waste, *Scientific American*, 29 December 2011, available at https://www.scientificamerican.com/article/webber-more-efficient-foods-less-waste/, accessed 7 September 2017.

11
TRAVEL

Robert Vale, Fabricio Chicca and Brenda Vale

Global transport trends

Between 1960 and 2014, the world's transport CO_2 emissions have represented around 20 per cent of total emissions, varying from a low of just under 19 per cent to a peak of just over 22 per cent over this period, with the 2014 figure being 20.5 per cent (The World Bank, 2017), making it a major contributor to increasing CO_2 in the atmosphere. Transport uses nearly two-thirds of the world's oil supply as its fuel. In 1973, the transportation sector worldwide used 45 per cent of oil production, with around a further 20 per cent used by industries, but by 2012, the share of oil demand for transportation had increased to 64 per cent (IEA, 2014: 33). In the European Union, transport is the only sector of the economy whose greenhouse gas (GHG) emissions have continued to increase since 1990, rising 14 per cent by 2012 (UIC/CER, 2015: 4).

Personal transport: Cars

Transport represented 23.3 per cent of global CO_2 emissions in 2005, and road transport formed by far the largest part of this at 17 per cent, with the remainder divided between shipping, aviation and all other modes (OECD/ITF, 2010: 7). In wealthy countries, two-thirds of these road-related CO_2 emissions are from passenger vehicles (OECD/ITF, 2010: 9), which is why personal choices are important. There is good information available to allow cars to be compared from the perspective of environmental impact. The US Department of Energy publishes a list of fuel consumption data, including "best" and "worst" figures (US Department of Energy, 2017a). Out of conventional petrol cars, the best in mid-2017 have a combined city/highway fuel consumption of 6.7 litres per 100 kilometres, while the worst use three times more fuel, 19.6 litres/100 kilometres.

Better performance is offered by petrol/electric hybrid cars, the best of which use 4.1 litres/100 kilometres.

The US Department of Energy also publishes equivalent figures for battery electric cars (US Department of Energy, 2017b), which, unlike hybrids, use electricity as their sole fuel source. The figures give a miles-per-gallon equivalent to show how far a car could go on a quantity of fuel with the same energy content as a US gallon of gasoline (US Department of Energy, 2017c). The combined figure for the best electric car works out at 1.73 litres/100 kilometres and for the least best at 3.27 litres/100 kilometres. In 2017, the best gasoline (hybrid) car and the best electric car available in the United States were both versions of the Korean Hyundai Ioniq. The figures reveal that it is simplistic to say "electric cars good, petrol cars bad". They show that battery electric cars do go further on a given amount of energy than conventional cars. The worst electric car is twice as good as the best petrol cars, and the best is four times as good.

What the figures do not reveal is the emissions related to electric cars, which depend on the electricity used to charge them. In the United States, coal provides 96 per cent of electricity in West Virginia and 0 per cent in Idaho and Rhode Island (Nunez, 2015). The US Department of Energy (2017d) has a website for comparing vehicle emissions in different states based on the mix of energy sources that each state uses to generate its electricity. In West Virginia, with nearly all its electricity coming from coal, the lowest emission option is a gasoline hybrid car with annual emissions of 2.8 tonnes compared with 5.2 for a conventional car and 4.3 for a battery electric car. In Rhode Island, which gets nearly all its electricity from natural gas, the figures are quite different, with the battery car having 1.9 tonnes of emissions compared with the hybrid and conventional car, as before. In the other non-coal state, Idaho, 59 per cent of electricity comes from hydro and the battery car emits only 0.4 tonnes (all data from US Department of Energy, 2017d). A Norwegian study found similar results, with a Toyota Prius hybrid having similar emissions to an electric Nissan Leaf charged using electricity with the global average composition of 40 per cent from coal, 25 per cent from gas, 5 per cent from oil and the remaining 30 per cent from zero-CO_2 sources (Holtsmark and Skonhoft, 2014).

Whether an electric car reduces emissions will depend not only on which car you buy, but also where you plug it in. Electric cars bring emissions reductions according to the percentage of renewable energy in the electricity grid. Even then, they may not be beneficial. Although 98 per cent of Norway's electricity comes from renewable sources (Ministry of Petroleum and Energy, 2016), the promotion of electric cars has resulted in more two-car households and more people giving up public transport in favour of driving, and the only long-term solution to reducing transport emissions is to bring in policies that restrict all car use (Holtsmark and Skonhoft, 2014).

Personal transport: Cycling

Individual transport does not mean only cars. Bicycles are widely promoted as a mode of transport (Dufour, 2010; Queensland Government, 2017). Cycling uses more energy on the part of the rider than sitting in a car or on the bus, so to allow a fair comparison with other modes, the additional food eaten by the cyclist as "fuel" needs to be taken into account. A study by the European Cyclists' Federation gives figures for the production and operation of a bicycle, a "pedelec" (an electrically assisted bicycle that the rider had to pedal as well), an average-sized European car and a bus with ten passengers. Table 11.1 shows the emissions for the different vehicles in grams of CO_2 equivalent per passenger-kilometre (Blondel et al., 2011: 9–15).

The food for cycling is assumed to be the average EU diet; if the cyclist were fuelled by eating beef, the emission would be 157 grams of CO_2 equivalent per kilometre (Blondel et al., 2011: 11). The car is assumed to be an average car with an average mix of driving and an average lifespan. The electricity for charging the pedelec assumes the EU average mix of generation sources, not a high percentage of coal. The problem with such calculations is that they have to be based on averages and reasonable assumptions, and the findings may be transferred to situations where there may be unintended consequences. Just as electric cars may not be preferable in some situations to petrol cars, so cycling may not be as much of a low-emissions choice of transport as its proponents might hope. But, on average, cycling, whether electrically-assisted or not, is more than ten times better in emissions terms than driving a car, and five times better than getting the bus.

Public transport

Probably the most common public transport is the bus. The US Department of Energy (2015) gives an average for "transit buses" of around 72 litres per 100 kilometres, roughly ten times that of a typical car. The problem with public transport vehicles is that fuel consumption per passenger-kilometre will vary depending on how many passengers are on the bus. When a bus sets out from one end of its route in the suburbs, it may have only a few passengers, but it will be full by the time it gets into the city. With fuel consumption of 72 litres per 100 kilometres, a bus would need to have ten passengers in order to be as fuel efficient as a typical car. Depending on the time of day and the number of people on the bus, it is not necessarily more fuel efficient to use public transport. However, the bus runs during the day, whether or not it has passengers, so each additional passenger will improve its average fuel consumption. When the bus is full, the situation improves

TABLE 11.1 Emissions per passenger-kilometre for bicycles, car and bus based on data from Blondel et al., 2011

Vehicle	Manufacture	Food	Fuel	Total
Bicycle	5	16		21 grams CO_{2e}
Pedelec	7	6	9	22 grams CO_{2e}
Car	42		229	271 grams CO_{2e}
Bus	6		95	101 grams CO_{2e}

considerably. A typical bus with all seats full and some standing passengers will have 54 to 64 passengers (Transportation Research Board, 1998: 2–26). With a full load of passengers, the fuel consumption of a city bus per passenger might be between 1.1 and 1.3 litres per 100 kilometres, which is much better than a single person in a car. On the basis of diesel fuel having a CO_2 content of 2.68 kilograms per litre (Davies, n.d.), the fuel consumption emissions of a rush hour bus will be around 29 to 35 grams per passenger-kilometre. This ignores the emissions of manufacturing, as discussed in the section on cycling, but the figure matches reasonably well with the European Cyclists' Federation figure of 95 grams per passenger-kilometre for the fuel emissions of a bus with ten passengers.

The problem of defining the emissions from transport modes is made clear in a recent report (M. J. Bradley and Associates, 2014) based on the amount of fuel purchased and the annual passenger miles for each operator of a particular mode, so these are figures showing the reality of the situation. The figures have been converted here in Table 11.2 to grams of CO_2 per passenger-kilometre, and organised into categories. Whether a mode is powered by electricity is also shown, to see if this might explain some of the differences.

The modes in Table 11.2 are ordered in terms of their average emission from lowest to highest. What is surprising is how extreme the differences between the

TABLE 11.2 CO_2 emissions in grams per passenger-kilometre for a range of urban and long-distance transport modes in the United States, based on M. J. Bradley and Associates, 2014

Mode	Electric?	Low	Ave	High
Urban public transport				
Van pool	No	31	**61**	106
Heavy rail	Yes	62	**79**	276
Transit bus	No	25	**85**	1,729
Commuter rail	Partly	89	**114**	200
Trolleybus	Yes	121	**143**	276
Light rail	Yes	73	**165**	2,444
Ferry boat	No	237	**512**	3,290
Urban private transport				
Car pool, two people	No	53*	**115***	177*
Car average trip	No	76*	**166***	254*
Car, one person	No	106*	**230***	354*
Long distance public				
Motorcoach	No	26	**27**	29
Intercity rail (AMTRAK)	Partly	88	**92**	101
Domestic air travel	No	–	**118**	–
Long distance private				
Car average trip	No	76*	**166***	254*
Car, one person	No	106*	**230***	354*

*Low assumes a hybrid car, average assumes an average US car and high assumes an SUV

TABLE 11.3 CO_2 equivalent figures per passenger-kilometre (from Fig. 1.6, IEA, 2009: 52)

Mode	CO_{2e} grams per passenger-kilometre	Range of values
Rail	40	10–50
Bus	45	25–80
Two wheelers	50	25–120
Cars	180	80–280
Air	230	225–255

low and high values in some cases are, particularly the transit bus, light rail and ferry boat. Given the difference between the average and high values, it is likely that these extreme figures are outliers rather than representative values, so it is probably reasonable to use the average values. The figures in Table 11.2 can be compared with more general figures as shown in Table 11.3.

What is again clear is that there is a great distance between the best and the worst in the case of some modes, particularly the private ones of two-wheelers and cars.

Limits to growth?

Table 11.2, which is based on measured data from many transport operators in the United States, shows that public transport users might have emissions that range between 61 grams of CO_2 per kilometre and 512 grams, depending on what mode they choose. Car users might have a range between 106 and 354 grams. What all these transport figures do not tell us is whether any of these modes are appropriate for a climate-constrained future. As far as transport emissions are concerned, how low is low enough? The city of Wellington, the capital of New Zealand, has the stated aim to reduce its emissions by 80 per cent compared with 2001 (WCC, n.d.). A recent study was made of how transport in Wellington would need to change to achieve such a reduction in line with the New Zealand government signing up to the Paris Climate Agreement, and serves as an example of what any city will need to do in the future. Given the need to reduce emissions by at least 80 per cent, combined with the likely growth in population by 2050, if individuals in the Wellington study wanted to have the same annual travel distance in 2050 that they have at present, the fuel-related emissions from any mode of transport, public or private, that they use to achieve their annual travel would have to be no more than 30 grams of CO_2 per passenger-kilometre (Shaikh and Vale, 2017). This is a very severe restriction; it rules out all the modes of transport in Table 11.2 except for the long-distance coach. It would, however, permit the use of bicycles, either electric or pedal, as shown in Table 11.1. This reveals the immense gap between the systems of transport that are currently used in modern cities and the systems of transport that will be needed if severe climate change is to be avoided.

The limit figure of no more than 30 grams of CO_2 per kilometre has considerable implications for the idea of the car as the main mode of travel in cities. Average CO_2 emissions of light vehicles entering the fleet in New Zealand (Ministry of

Transport, 2016: 55) show a gradual reduction over 10 years from 225 grams/kilometre in 2005 to less than 170 grams/kilometre in 2015. This is a reduction of 5.5 grams a year for vehicles joining the vehicle fleet, but the figure is still nearly six times the limit figure of 30 grams/kilometre. The rate of change has been slow, so it is unlikely that average vehicles will quickly become more efficient in their use of fuel. The average age of cars leaving the New Zealand light vehicle fleet was 19 years in 2015; this has risen steadily since 2007 (Ministry of Transport, 2016: 41). If the trend continues, it is possible that conventional vehicle technology could be improved, but not greatly, giving the likelihood of average fleet emissions of less than 100 grams/kilometre by 2050.

One hundred grams/kilometre is nowhere near the limit of technology. Volkswagen has already marketed the world's most fuel-efficient car, a diesel-electric hybrid two-seater that achieves a fuel consumption of 110 kilometres/litre (0.91 litres/100 kilometres) (VW, 2013), which is 24 grams/kilometre, but such performance will not represent the fleet average by 2050. If the "limit" is 30 grams per kilometre and the best likely to be achieved by 2050 as a fleet average is 100 grams per kilometre, car travel will have to be limited to a third of the current distance, which is a huge change. As Table 11.2 shows, none of the current modes of public transport, apart from the long distance coach, falls within the limit either.

Perhaps the answer is electric cars, which are more efficient in their use of energy than oil-burning cars. Electric vehicles have a typical CO_2 emissions level of 50 grams/kilometre (Manjunath and Gross, 2017; Jochem et al., 2015), which could give an incentive to shift to electric cars, but even changing to electric cars would mean a reduction in annual travel distance as this is nearly twice the limiting value of 30 grams per kilometre.

Self-driving vehicles are not likely to have lower fuel consumption and are not likely to reduce travel distances; they are more likely to encourage greater movement. If avoiding climate change is the goal rather than just making money, it would be better if the effort going into robotic vehicles were to go into fuel efficiency. In 1958, a Swiss guide to world cars listed twenty models with fuel consumption below 5 litres per 100 kilometres, with one as low as 2.9 (International Automobile Parade, 1958). There has been little or no progress in the last 60 years in terms of improving the average fuel consumption of the world's car fleet.

Freight transport

Increasingly, goods move long distances. Just as different modes of passenger transport have different rates of emissions, the same applies to freight, with the unit of measurement being a tonne of freight moved one kilometre. Table 11.4 shows the range.

The shock in Table 11.4 is the figures for vans. The average emissions for a van mean that if you import something, the emissions produced in transporting it 7,000 kilometres on a ship could be the same as transporting it 100 kilometres in a van from the port of arrival. Commoner (1972) pointed out that shifting freight by rail

TABLE 11.4 CO_2 emissions for freight in grams per tonne-kilometre

Mode	Emission in grams CO_2 per tonne-kilometre
Very large container vessel	3*
Maritime (average)	14**
Rail (electric)	18**
Rail (average)	21**
Rail (diesel)	29**
Inland shipping	61**
Road (average)	75**
40-tonne truck	80*
Van (diesel)	397***
Boeing 747	435*
Van (diesel)	572****
Van (diesel)	752*****
Van (diesel)	2240******

*(ICS, 2014)
**European data for 2011 from EEA (2015)
***(ECMT, 2007)
****(Steenhof et al., 2006)
*****(Perez-Martinez and Sorba, 2010)
******(Lenzen, 1999)

in the United States used six times less energy than sending it by truck, but despite this advantage, in the United States rail freight haulage had been displaced by trucking since 1946. This was good for the "free market" and the road transport industry, but not necessarily for the environment.

What about biofuels?

If transport fuels could be made from plants, they could be argued to be carbon-neutral because the carbon released by their combustion would have been taken from the atmosphere by the growing plants. If liquid fuels could be made from plants, they could allow the use of existing cars, aircraft and other transport modes. By 2050, biofuels could be providing 27 per cent of total transport fuels at little or no additional cost while avoiding 2.1 gigatonnes of CO_2 emissions annually (IEA, 2011: 5). It is also argued that there is no conflict between using land for growing biofuels and using land for growing food as the average area of cropland used to feed one person fell by half from 1961 to 2006 and is now thought to be around 0.19 hectares. This is the land physically occupied by crops rather than the total ecological footprint. If it takes 0.19 hectares to feed one person, it would require 1.7 billion hectares to feed the projected world population of 9 billion in 2050. This is a third of the current area of arable land (Kline et al., 2017), so there would appear to be plenty of land left to grow fuels.

The IEA (2011: 16) shows that biofuels offer a broad range of GHG savings compared with oil. Some of the "conventional" biofuels (those which are already in production), such as ethanol made from sugarcane, show a GHG saving of 75 to 100 per cent compared with oil, while others, including some of those labelled "advanced", such as biodiesel made from algae, have the potential to have emissions 50 per cent greater than conventional fuels.

Searchinger et al. (2008) suggest that studies claiming biofuels have a positive effect on reducing carbon emissions have failed to take account of carbon emitted as a result of farmers responding to market signals. If more crops are grown for fuel, this will cause the price of food to rise; higher prices can be earned for food crops and this encourages farmers to convert land which is currently forest or grassland into new cropland to replace the land whose crops are being converted into biofuels instead of being used for food. Alternatively, farmers may bring additional land into cultivation in order to grow fuel crops. Either way, the existing carbon sequestration effect of the land is changed. If land is being cropped, the carbon that the plants take out of the air is not stored in the plants but gets back into the air after the crops are eaten. Eventually, there may be a benefit as the carbon saving due to using biofuel instead of oil gradually covers the carbon sequestration lost by converting wild land into cropland, but this can take a long time. Producing ethanol from maize in the United States reduces emissions by 20 per cent compared with using oil if the land-use change is not taken into account, but with the land-use changes there is an increase of emissions for the next 167 years (Searchinger et al., 2008). Once again, there is "no such thing as a free lunch", as Barry Commoner argued nearly 50 years ago (Commoner, 1972).

Working it out

To demonstrate how the figures were calculated, transport details for some randomly chosen stories are analysed below. In all cases, the assumptions made are such that the overall distance and fuel consumption are likely to be under- rather than over-estimated.

Transport example 1: Johannesburg, South Africa—a one-car household

The car for this story is assumed to be a Renault Sandero Stepway with a 1.6-litre petrol engine, built in 2010. Fuel consumption for this 2010 model is 7.2 litres/100 kilometres for combined urban and highway driving (Davids, 2010). A recent study of European fuel consumption ratings found an increasing discrepancy between official fuel consumption data and fuel consumption achieved in use (Mock et al., 2014: 45). On this basis, the average fuel consumption of this car could be closer to 10 litres/100 kilometres. *Consumer Reports* (2013) in the United States found that the situation had improved over time. The difference between the European and US findings is likely to be due to the different testing procedures that are used.

Fuel consumption could also be greater than expected, given that the urban consumption of cars is always greater than the extra urban/highway consumption

(Vale and Vale, 2013: 68), and that most of the travel in this case is urban. However, in order to ensure that the results cannot be considered exaggerated, this analysis uses published combined fuel consumption figures.

The distances travelled are as follows:

- School, work and back—assume 20-kilometre round trip, 12 kilometres to Dora's work and 8 kilometres to school.
- Weekend Church 2-kilometre round trip.
- Weekend shopping—assume 10-kilometre round trip (say, 15 kilometres per weekend).
- Yearly work and school 100 kilometres per week for 50 weeks = 5,000 kilometres.
- Church and shopping 52 weeks × 15 kilometres = 780 kilometres.
- Say, 6,000 kilometres total per year. At 7.2 litres/100 kilometres, this will be 432 litres of petrol per year.

Transport example 2: Tucumán, Argentina—a two-car household with some bus use

There are two cars in this household. One is assumed here to be a Volkswagen Suran Cross with a 1.6-litre petrol engine, built in 2015, giving it a combined urban/highway fuel consumption of 7.5 litres per 100 kilometres. (Anon, 2015). The other car is assumed to be a Volkswagen Vento with the same fuel consumption as the Suran Cross (Anon, n.d. a). Since both work in the same area in the city, the couple travel approximately the same distance each month.

Work is roughly 12 kilometres from home, making it a 25-kilometre round trip.

Work travel

- Mon, Tue, Wed: 3 × 25 kilometres = 75 kilometres in the Suran = 300 kilometres per month.
- Thu, Fri: 2 × 25 kilometres = 50 km in the Vento = 200 kilometres per month.
- Bus every day, one way = 12.5 kilometres × 5 days = 62.5 kilometres = 250 kilometres per month.

Sharing a car trip means that only one car makes the trip to work and back, although the car comes back with only one occupant and the other person takes the bus. The two cars travel a total of 12,000 kilometres in a year. Added to this is the annual holiday at the beach, a round trip of 3,000 kilometres. This gives a total car use of 15,000 kilometres per year for this household, and for this they use 1,125 litres of petrol a year.

One or other of the members of this household also travels by bus for the trip home from work each day, making 250 kilometres a month or 3,000 kilometres of bus travel per year. Based on these figures, and assuming that the bus travel of this household takes place at a time when the bus is full, bus travel represents an additional 36 litres of diesel per year. So this household's annual fuel consumption for transport is 1,125 litres of petrol and 36 litres of diesel.

Transport example 3: São Paulo, Brazil—a family with two cars that can use two different fuels and some additional bus travel

The family have two cars, which are assumed here to be a Volkswagen Golf 1.6 Flex and a Fiat Uno 1.0 Flex. The "Flex" after the car's name indicates that it can be fuelled with petrol or sugar cane ethanol. This happens according to the seasons, as the price of ethanol changes over the year. If the ethanol is not 30 per cent cheaper, the family use petrol, as it is the more efficient fuel. The family have two cars despite only using one during the week. This is because the city imposes a car circulation restriction once a week according to the car's number plate.

Fuel consumption for the VW (combined) is 14.4 kilometres/litre on petrol and 9.5 kilometres/litre on alcohol (Anon, 2016a), which equals 6.9 litres per 100 kilometres on petrol and 10.5 litres per 100 kilometres on alcohol. The fuel consumption for the Fiat is slightly more complicated as the published figures say that it manages 14.3 kilometres/litre urban and 18.2 kilometres/litre rural on petrol, and 10.1 kilometres/litre urban and 12.4 kilometres/litre rural when running on ethanol (Anon, 2016b). A combined value is not given. Applying the protocol used in Brazil for calculating the combined value (Yang et al., 2015: 40; US Department of Energy, n.d. a) to the figures for the Fiat gives 6.2 litres per 100 kilometres on petrol and 9.0 litres per 100 kilometres on alcohol.

> VW Golf Flex: 6.9 litres /100 kilometres on petrol and 10.5 litres/100 kilometres on ethanol
> Fiat Uno Flex: 6.2 litres/100 kilometres on petrol and 9.0 litres/100 kilometres on ethanol

The use of ethanol makes the calculation more complicated. The family drive a total of 110 kilometres a week for work and school plus an assumed further 25 kilometres at the weekend, giving an annual distance of around 7,000 kilometres. Again assuming that they use ethanol for half the time, each car will run 26 weeks of the year on ethanol and 26 weeks on petrol, making 1,750 kilometres for each car on each fuel. This means that in a year, the VW will use 121 litres of petrol and 184 litres of ethanol, and the Fiat will use 109 litres of petrol and 158 litres of ethanol.

The household's total annual consumption for their car travel will be 230 litres of petrol and 342 litres of ethanol.

In addition, each month there is 75 kilometres of bus travel. Assuming this is the same each month means an additional 900 kilometres of bus travel. Using the figure for a full bus from transport example 2, this bus travel will use a further 11 litres of diesel.

Transport example 4: Dourados, Brazil—a family that does a lot of driving

This farming family have eight vehicles, of which four (all four-wheel drive) are mostly dedicated to work, two are used for the family routine, and two for leisure. Respectively, for the calculations, the vehicles are assumed to be:

Work vehicles	
Land Rover Evoque 4 × 4,	8.4 litres/100 kilometres (1)
Porsche Cayenne 4 × 4,	10.7 litres/100 kilometres (2)
Ford F 4000 diesel 4 × 4,	11.6 litres/100 kilometres (3)
Toyota Hilux diesel 4 × 4	8.9 litres/100 kilometres (4)
Family vehicles	
Nissan Sentra	7.8 litres/100 kilometres (5)
Audi A5	7.8 litres/100 kilometres (6)
Leisure vehicles	
Maserati Quattroporte	13.1 litres/100 kilometres (7)
Porsche Cayman	9.0 litres/100 kilometres (8)

1. Land Rover Evoque: The US EPA figures for this are 19 miles per gallon urban and 28 miles per gallon highway (Swan, 2011). Using the 55 per cent urban/45 per cent highway split (US Department of Energy, n.d. b), this gives a combined figure of 23.05 miles per gallon (US), which is 10.2 litres/100 kilometres, but it might be fairer to use only the highway value as much of the driving is long distance, which would give this car a figure of 8.4 litres/100 kilometres.
2. Porsche Cayenne: The fuel consumption figures are as follows: urban 22.1 litres/100 kilometres, extra urban 10.7 litres/100 kilometres, combined 14.9 litres/100 kilometres (Anon, n.d. b). Again, the extra urban figure appears to be the more appropriate one to use.
3. Ford F-4000: Because the Ford F-4000 is supplied as a cab chassis on which a body is then constructed, there are no fuel consumption figures published by the manufacturer. The larger-engined Ford F-250 XLT 3.9 Turbo 4×4 CS 2010 Diesel, which comes with full bodywork, has figures of 6.7 kilometres/litre urban and 8.8 kilometres/litre rural (Consumo Combustível, 2013a), which work out at 14.9 litres/100 kilometres urban and 11.6 litres/100 kilometres rural. The rural figure will be used here.
4. Toyota Hilux: The fuel consumption of a non-automatic version is given as 9.5 kilometres/litre urban and 11.2 kilometres/litre rural (Consumo Combustível, 2013b). These work out at 10.5 litres per 100 kilometres urban and 8.9 litres per 100 kilometres rural. The rural figure will be used.
5. Nissan Sentra: This has a fuel consumption of 10.2 kilometres/litre urban and 12.9 kilometres/litre rural when using petrol (iCarros, n.d. a). This works out at 9.8 litres/100 kilometres urban and 7.8 litres/100 kilometres rural. Although it is a "flex" vehicle, like the examples from São Paulo, the family did not report using ethanol, so it will be assumed to be running on petrol with a consumption of 7.8 litres/100 kilometres.
6. Audi A5: The fuel consumption is 9.0 kilometres/litre urban and 12.9 kilometres/litre rural (iCarros, n.d. b). These figures work out at 11.1 litres /100 kilometres urban and 7.8 litres/100 kilometres rural. The rural figure will be used here.

7. Maserati Quattroporte: The fuel consumption on the European urban cycle is given as 9 miles per gallon, on the extra-urban cycle 18 miles per gallon and the combined cycle 13 miles per gallon (Robinson, 2004). These figures convert to 26.1 litres/100 kilometres urban, 13.1 litres/100 kilometres extra urban and 18.1 combined. The extra urban figure will be used here.
8. Porsche Cayman: The fuel consumption is 22 miles per gallon combined city/highway, 19 miles per gallon city, 26 miles per gallon highway (US Department of Energy, n.d. b). These numbers work out at 10.7 litres /100 kilometres combined, 12.4 litres/100 kilometres city and 9.0 litres/100 kilometres highway.

The figures in the story suggest the following yearly distances:

Home to school and back (children and driver)	16,000 kilometres
Home to soy farm and back (Pedro)	18,700 kilometres
Trips to town (Pedro)	7,200 kilometres
Trips to town (Gabriele)	11,700 kilometres
Total distance	53,600 kilometres (1,031 kilometres a week)

The trips recorded in the week of the study add up to 1,139 kilometres, which accords well with the estimate of about 1,031 kilometres per week. The Ford four-wheel drive and the Maserati and Porsche sports cars were not used during this week; the Ford and the Maserati have the highest fuel consumption out of the eight vehicles.

Day	Person	Vehicle	Distance	Consumption	Total fuel
Monday	Pedro	Hilux	55 kilometres	8.9 litres/100 kilometres	4.9 litres
	Gabriele	Audi	48 kilometres	7.8 litres/100 kilometres	3.7 litres
	Driver	Nissan	95 kilometres	7.8 litres/100 kilometres	7.4 litres
Tuesday	Driver	Nissan	61 kilometres	7.8 litres/100 kilometres	4.8 litres
Wednesday	Pedro	Land Rover	180 kilometres	8.4 litres/100 kilometres	15.1 litres
	Gabriele	Audi	58 kilometres	7.8 litres/100 kilometres	4.5 litres
	Driver	Nissan	88 kilometres	7.8 litres/100 kilometres	6.9 litres
Thursday	Pedro	Land Rover	180 kilometres	8.4 litres/100 kilometres	15.1 litres
	Gabriele	Audi	51 kilometres	7.8 litres/100 kilometres	4.0 litres
	Driver	Nissan	68 kilometres	7.8 litres/100 kilometres	5.3 litres

(Continued)

Day	Person	Vehicle	Distance	Consumption	Total fuel
Friday	Gabriele	Audi	77 kilometres	7.8 litres/ 100 kilometres	6.0 litres
	Driver	Nissan	65 kilometres	7.8 litres/ 100 kilometres	5.1 litres
	Driver	Cayenne	42 kilometres	10.7 litres/ 100 kilometres	4.5 litres
Saturday	Gabriele	Cayenne	71 kilometres	10.7 litres/ 100 kilometres	7.6 litres
Sunday	At home all day				nil
Weekly total			1,139 kilometres	petrol	90 litres
				diesel	4.9 litres
Yearly total (all fuels)			59,228 kilometres		5,030 litres
Average fuel consumption					8.3 litres/100 kilometres

The figures show that in a year, the family will be using at least 5,030 litres of fuel. In the week in question, only a small part of this was diesel for their one diesel vehicle, so for simplicity, their diesel consumption has not been separated out from their total annual fuel consumption. The average fuel consumption of their travel over the week is equivalent to a car that uses 8.3 litres/100 kilometres, not a particularly high figure. It is not so much the number or the type of cars that they own that leads to high consumption, it is the distances that they travel. Of their travel, 35 per cent of the distance, 18,700 kilometres per year, is work-related, based on the need to travel to and from the distant soy farm. Even without this, this family still have a high travel demand as an inevitable result of needing to live out of the town in order to be able to operate as farmers, but at the same time needing to access the services of the town. Could they reduce their impact? Certainly, they could consider using more fuel-efficient vehicles, and what is also interesting is that their daily travel distances in the week under study, undertaken in five of their eight vehicles, all fall comfortably within the range of currently available electric vehicles.

Comparison

So how do the travel arrangements of the people in the stories compare with one another? Table 11.5 shows the total annual travel distance for each household, the fuel that they use for this and the number of people for whom the travel data apply.

The range of carbon dioxide emissions among our families is very large, especially considering that these emissions do not take into account any international air travel. The best-performing families have almost zero emissions; the highest emissions for one family are over 14 tonnes a year. If we list the emissions per person, to take account of different family sizes, the result is as shown in Table 11.6,

TABLE 11.5 CO_2 emissions related to transport for each of the stories. Bold numbers indicate totals. The final figures assume that the multiplier for the embodied energy for extraction, refining and shipping of petroleum products is 1.25 (Alcorn, 2001: 7)

Country	No/Location	Vehicle	Distance	Fuel	Quantity	CO_2-factor	Emission kg CO_2/year	
Morocco	4 urban	walking	3,500	mixed		58 grams/kilometre (1)		**203**
Mozambique	10 rural	boat	15,000	wind				nil
		walking	21,840	vegetables		2 grams/kilometre		44
		Total	36,840				Total	**44**
South Africa	5 urban	car	6,000	petrol	432 litres	2.31 kilograms/litre (2)	998	**1,248**
Argentina	4 urban	two cars	15,000	petrol	1,125 litres	2.31 kilograms/litre	2,599	3,249
		bus	3,000	diesel	36 litres	2.68 kilograms/litre (2)	96	120
		Total	18,000				Total	**3,369**
Brazil	4 urban	two cars	7,000	petrol	230 litres	2.31 kilograms/litre	531	664
				ethanol	342 litres	0.23 kilograms/litre (3)	79	99
		bus	900	diesel	11 litres	2.68 kilograms/litre	29	36
		Total	7,900				Total	**799**
Brazil	5 rural	eight cars	53,600	petrol	4,680 litres	2.31 kilograms/litre	10,811	13,514
				diesel	255 litres	2.68 kilograms/litre	683	854
		Total	53,600				Total	**14,368**
South Sudan	13 rural	bicycle	10,000	vegetables		2 grams/kilometre (1)	20	
		bus	10,000	diesel	120 litres	2.68 kilograms/litre	322	403
		Total	20,000				Total	**423**
South Sudan	6 UNMISS	motorbike taxi (a)	2,300	petrol	70 litres	2.31 kilograms/litre	162	**203**
India	3 urban	walking	6,000	vegetables		8 grams/kilometre		48
Indonesia	5 rural	walking	16,000	vegetables		8 grams/kilometre		128
Japan	5 urban	car	2,000	petrol	150 litres	2.31 kilograms/litre	347	434
		motorcycle (a)	4,500	petrol	135 litres	2.31 kilograms/litre	312	390
		Total	6,500				Total	**824**

Malaysia	6 urban	two cars	20,000	petrol	1,893 litres	2.31 kilograms/litre	4,373	5,466
Mongolia	3 urban	car	2,100	petrol	153 litres	2.31 kilograms/litre	353	441
		bus/van	3,500	diesel (b)	42 litres	2.68 kilograms/litre	113	141
		Total	5,600				Total	582
Myanmar	4 rural	bicycle	8,000	vegetables		2 grams/km		16
		walking	1,000	vegetables		8 grams/km		8
		boat	9,000	petrol	1,560 litres	2.31 kilograms/litre	3,604	4,505
		Total	18,000				Total	4,529
Finland	4 urban	car	10,760	diesel	484 litres	2.68 kilograms/litre	1,297	1,621
		motorcycle	5,400	petrol	275 litres	2.31 kilograms/litre	635	794
		bicycle	2,000	mixed		16 grams/km		32
		Total	18,160				Total	2,447
Germany	2 village	car	30,000	diesel	1,500 litres	2.68 kilograms/litre	4,020	5,025
		train	1,400	electricity (4)		35 g/passenger-kilometre		49
		Total	31,400				Total	5,074
England	3 urban	car	13,000	petrol	936 litres	2.31 kilograms/litre	2,162	2,703
		train	13,000	electricity (4)		55 g/passenger-kilometre		715
		walking	850	mixed		58 grams/kilometre		49
		Total	26,850				Total	3,467
Canada	2 urban	car	19,500	petrol	1,989 litres	2.31 kilograms/litre	4,595	5,744
		streetcar	500	electricity (5)		55 g/passenger-kilometre		28
		walking	420	mixed		58 grams/kilometre		24
		Total	20,420				Total	5,796
Cuba	3 urban	bus	2,600	diesel	31 litres	2.68 kilograms/litre	83	104
		walking	420	mixed		58 grams/kilometre		24
		Total	3,020				Total	128
United States	2 rural	two cars	28,600	petrol	3,203 litres	2.31 kilograms/litre	7,399	9,249
		side-by-side	400	diesel	40 litres	2.68 kilograms/litre	107	134
		Total	29,000				Total	9,383

(Continued)

TABLE 11.5 (Continued)

Country	No/Location	Vehicle	Distance	Fuel	Quantity	CO_2 factor	Emission kg CO_2/year
United States	4 urban	two cars	25,000	petrol	1,500 litres	2.31 kilograms/litre	3,465
New Zealand	4 rural	two cars	13,300	petrol	1,254 litres	2.31 kilograms/litre	2,897
		quad bike	3,100	petrol	214 litres	2.31 kilograms/litre	494
		Total	16,400				**Total 4,331**
							3,621
							618
							4,239
Australia	3 urban	two cars	2,000	petrol	81 litres	2.31 kilograms/litre	187
		train	2,500	diesel	43 litres	2.68 kilograms/litre	115
				electricity		45 g/passenger-kilometres (6)	113
		walking	850	mixed		58 grams/kilometre	49
		Total	5,350				**Total 540**
							234
							144
Tonga	4 urban	bus	8,200	diesel	98 litres	2.68 kilograms/litre	263
		walking	900	mixed		58 grams/kilometre	52
		Total	9,100				**Total 381**
							329

(a) Based on the consumption of a small motorbike being 3.0 litres per 100 kilometres, using data from TMW (n.d.) *2010 Motorcycle Fuel Economy Guide*, available at http://www.totalmotorcycle.com/MotorcycleFuelEconomyGuide/2010b-MPG?d=1, accessed 23 August 2017.
(b) Assuming that a bus and a shared van have similar emissions as shown in Table 11.2.

(1) Calculated from food and cycling data in Blondel et al. (2011: 10–11) and walking energy data from Vale and Vale (2009: 103).
(2) Data from Davies (n.d.).
(3) 10 per cent of the value for petrol based on IEA (2007: 1, 4).
(4) Data from IEA (2012: 39, Fig. 33) giving a value of 35 g/passenger-kilometre for intercity and high-speed trains and 55 g/passenger-kilometre for regional trains.
(5) Value from Table 2 for light rail, using one-third of the average value to account for commuter travel.
(6) Value for commuter trains in Australia from TTF (2009: 8).

Alcorn A (2001) *Embodied energy and CO_2 coefficients for New Zealand building materials* Wellington: Victoria University of Wellington, Centre for Building Performance Research.
IEA (2007) *Biofuel Production IEA Energy Technology Essentials* January. Paris: International Energy Agency.
IEA (2012) *Railway Handbook 2012* Paris: International Energy Agency.
TTF (2009) *Public Transport and Climate Change – TTF Transport Position Paper* November, Sydney: Tourism and Transport Forum.

TABLE 11.6 Carbon dioxide emissions for transport kilograms per person, annual distance travelled per person per year and emissions in grams per person-kilometre per year

Country	Urban or rural	CO_2 kilograms/ person	kilometres/ person	CO_2 grams/ person-kilometres
No car				
Mozambique	Rural	4	3,684	1
India	Urban	16	2,000	8
Indonesia	Rural	26	3,200	8
South Sudan	Rural	33	1,538	22
South Sudan	UNMISS	34	383	89
Cuba	Urban	43	1,007	43
Morocco	Urban	51	875	58
Tonga	Urban	95	2,275	42
Car owners—short distance				
★Japan	Urban	165	1,300	128
★Australia	Urban	180	833	217
★Mongolia	Urban	194	1,867	104
★Brazil	Urban	200	1,975	101
★South Africa	Urban	250	1,200	208
Car owners—medium distance				
★Finland	Urban	612	4,540	135
★Argentina	Urban	842	4,500	187
★Malaysia	Urban	911	3,333	273
★New Zealand	Rural	1,060	4,100	259
★United States	Urban	1,083	6,250	173
xMyanmar	Rural	1,132	4,500	252
Car owners—longer distance				
★England	Urban	1,156	8,950	129
★Germany	Urban	2,537	15,700	162
★Brazil	Rural	2,874	10,720	268
★Canada	Urban	2,898	10,210	284
★United States	Rural	4,692	14,500	324

Italics show families which do not use any oil for transport (not even for public transport)
★ shows families which use a car
x shows a family that use an outboard motor

which is ordered in terms of kilograms of CO_2 per person (the third column), from the lowest at 4 kilograms per person for the Mozambique family and the highest at over 4.5 tonnes per person for the rural family in the United States. Columns 4 and 5 show how many kilometres are travelled by each person in the family and the CO_2 emissions per person-kilometre of travel. This also shows which families do not use any oil for their travel and which have cars (or in one case, an outboard motor on their boat).

Table 11.6 gives some useful indications, again showing that transport is more complex than we might think. Looking at column 4, we can see for example that not owning a car does not necessarily mean that your travel will be curtailed. Several of the families with cars travelled much shorter distances per person per year than the families without cars. On the other hand, owning a car (or an outboard motor) guarantees that your emissions per kilometre travelled will be higher than those of non-car owning families. However, owning, and using, a car does not necessarily mean that your yearly travel emissions per person will be high. If you own a car but choose not to do a lot of driving, like the families in Japan, Australia, Mongolia, urban Brazil and South Africa, your emissions will be only around 20 per cent of those of the next group of car owners, who travel medium distances. The families which do a lot of travelling have the largest emissions, but the German family, in spite of having the most travel per person, manage to keep their emissions lower by having a smaller car and doing some of their travel by train. The same applies to the family in England, who commute to work on the train; both families have fairly low emissions per person-kilometre.

Perhaps the most positive aspect of all this is that it makes clear that we can make a difference in our environmental impact through how we choose to travel, and that there are many ways of reducing your transport emissions.

References

Anon (n.d. a) *Características técnicas: Volkswagen - Vento (1HX0) - 2.0 (115 Hp)*, available at http://www.auto-data.net/es/?f=showCar&car_id=8851, accessed 5 April 2017.

Anon (n.d. b) Technical specifications: Porsche - Cayenne (955) Facelift - 4.8 S (385 Hp), available at https://www.auto-data.net/en/?f=showCar&car_id=6718, accessed 12 June 2017.

Anon (2015) *Crítica: Volkswagen Suran 1.6 16v Highline*, available at http://autoblog.com.ar/2015/06/05/critica-volkswagen-suran-1-6-16v-highline/, accessed 5 April 2017.

Anon (2016a) *Golf 1.6 8V Total Flex Manual 4P - Ficha Técnica*, available at http://www.vrum.com.br/fichatecnica/Volkswagen/GOLF/2011/005252-3, accessed 10 April 2017.

Anon (2016b) Uno Way 1.0 8V Evo MPI Flex Manual 4P - Ficha Técnica, available at http://correiobraziliense.vrum.com.br/fichatecnica/Fiat/UNO/2011/001305-6, accessed 10 April 2017.

Blondel B, Mispelon C and Ferguson J (2011) *Cycle more often 2 cool down the Planet: Quantifying CO_2 savings of Cycling*, Brussels: European Cyclists' Federation.

Commoner B. (1972) *The Closing Circle: Nature, Man and Technology*, New York: Alfred A Knopf.

Consumer Reports (2013) Why you might not be getting the efficiency promised, *Consumer Reports* magazine: August, available at http://www.consumerreports.org/cro/magazine/2013/08/the-mpg-gap/index.htm, accessed 14 June 2017.

Consumo Combustível (2013a) Consumo F-250 XLT 3.9 Turbo 4×4 CS 2010, available at http://consumocombustivel.com.br/consumo-f-250-xlt-3-9-turbo-4x4-cs-2010/, accessed 12 June 2017.

Consumo Combustível (2013b) Consumo Toyota Hilux SRV 3.0 Turbo 4×4 AT CD 2012 Diesel, available at http://consumocombustivel.com.br/consumo-toyota-hilux-srv-3-0-turbo-4x4-at-cd-2012-diesel/, accessed 12 June 2017.

Davids S. (2010) "Renault's sporty(ish) Sandero , *Wheels*, 24, 11 November, available at http://www.wheels24.co.za/NewModels/Renaults-sportyish-Sandero-20101110, accessed 3 March 2017.

Davies T. (n.d.) *Calculation of CO_2 emissions*, available at https://people.exeter.ac.uk/TWDavies/energy_conversion/Calculation%20of%20CO2%20emissions%20from%20fuels.htm, accessed 16 August 2017.

Dufour D (2010) *PRESTO Cycling Policy Guide General Framework* Intelligent Energy – Europe Programme granted by the Executive Agency for Competitiveness and Innovation (EACI).

ECMT (2007) *Cutting transport CO_2 emissions: What progress?*, Paris: European Conference of Ministers of Transport, OECD.

EEA (2015) *Specific CO_2 emissions per tonne-km and per mode of transport in Europe, 1995-2011* 3, August, Copenhagen, European Environment Agency, available at https://www.eea.europa.eu/data-and-maps/figures/specific-co2-emissions-per-tonne-2 accessed 21 August 2017.

Holtsmark B and Skonhoft A (2014) The Norwegian support and subsidy policy of electric cars. Should it be adopted by other countries? *Environmental Science and Policy*, 42(October). pp 160–168.

iCarros (n.d. a) *Ficha técnica de Audi A5 2015*, available at http://www.icarros.com.br/audi/a5/2015/ficha-tecnica/22072, accessed 12 June 2017.

iCarros (n.d. b) *Nissan Sentra SV 2.0 16V CVT (Aut) (Flex) 2014*, available at http://www.icarros.com.br/nissan/sentra/2014/ficha-tecnica/15704, accessed 12 June 2017.

IEA (2007) Biofuel Production, IEA Energy Technology Essentials, Paris: International Energy Agency.

IEA (2012) *Railway Handbook 2012t*, Paris: International Energy Agency.

International Automobile Parade (1958) *International Automobile Parade Vol II*, Zurich Arthur Logoz.

Jochem P, Babrowski S and Fichtner W (2015) Assessing CO_2 emissions of electric vehicles in Germany in 2030 *Transportation Research Part A: Policy and Practice*, 78, pp. 68–83.

Kline K, Msangi S, Dale V, Woods J, Souza G, Osseweijer P, Clancy J, Hilbert J, Johnson F, McDonnell P and Mugera H (2017) Reconciling food security and bioenergy: priorities for action, *GCB Bioenergy*, 9, pp. 557–576.

Lenzen M (1999) Total requirements of energy and greenhouse gases for Australian transport, *Transportation Research Part D*, 4, pp. 265–290.

Manjunath A and Gross G (2017) Towards a meaningful metric for the quantification of GHG emissions of electric vehicles (EVs), *Energy Policy*, 102, pp. 423–429.

Ministry of Transport (2016) *Annual Fleet Statistics, 2015*, Wellington: Ministry of Transport.

Ministry of Petroleum and Energy (2016) *Renewable Energy Production in Norway* 11 May, available at https://www.regjeringen.no/en/topics/energy/renewable-energy/renewable-energy-production-in-norway/id2343462/, accessed 14 August 2017.

M. J. Bradley and Associates (2014) *Updated Comparison of Energy Use & CO_2 Emissions From Different Transportation Modes* Washington, DC, American Bus Association. April.

Mock P et al. (2014) *From Laboratory to Road; a 2014 Update of Official and "Real-World" Fuel Consumption and CO_2 Values for Passenger Cars in Europe*, September, Berlin: International Council on Clean Transport.

Nunez C (2015) See which states use coal the most as new climate rule is finalized, *National Geographic*, 11 August, available at http://news.nationalgeographic.com/energy/2015/08/080115-seven-most-coal-dependent-states/, accessed 14 August 2017.

OECD/ITF (2010) *Reducing Transport Greenhouse Gas Emissions: Trends & Data 2010*, Leipzig: Organisation for Economic Cooperation and Development International Transport Forum.

Perez-Martinez P J and Sorba I (2010) Energy consumption of passenger land transport modes, *Energy and Environment*, 21(6), pp. 577–600.

Queensland Government (2017) *Promoting cycling*, 9 May, Brisbane, Department of Transport and Main Roads, available at https://www.tmr.qld.gov.au/Travel-and-transport/Cycling/Bike-user-guide/Promoting-cycling.aspx V; https://www.tmr.qld.gov.au/Travel-and-transport/Cycling/Bike-user-guide/Promoting-cycling.aspx, accessed 16 August 2017.

Robinson P (2004) *2005 Maserati Quattroporte*, available at http://www.caranddriver.com/reviews/2005-maserati-quattroporte-first-drive-review-specs-page-2, accessed 12 June 2017.

Searchinger T, Heimlich R, Houghton R, Dong F, Elobeid A, Fabiosa J, Tokgoz S, Hayes D and Yu T (2008) Use of U.S. croplands for biofuels increases greenhouse gases through emissions from land-use change, *Science*, 319, pp. 1238–1240.

Shaikh R and Vale R (2017) Zero-emission transport for the 'Coolest Little Capital' - How transport in Wellington, New Zealand can change to meet the Paris Climate Agreement Poster presentation, *ITE/CITE 2017 Annual Meeting and Exhibit* Toronto: Institute of Transportation Engineers, 30 July to 2 August.

Steenhof P, Woudsma C, and Sparling E (2006) Greenhouse gas emissions and the surface transport of freight in Canada, *Transportation Research Part D*, 11, pp. 369–376.

Swan T (2011) 2012 Land Rover Range Rover Evoque *Car and Driver*, August, available at http://www.caranddriver.com/reviews/2012-land-rover-range-rover-evoque-first-drive-review, accessed 12 June 2017.

The World Bank (2017) *CO_2 emissions from transport (% of total fuel combustion)*, available at http://data.worldbank.org/indicator/EN.CO2.TRAN.ZS, accessed 28 July 2017.

Transportation Research Board (1998) *Transit capacity and quality of service manual Part 2 Bus transit capacity*, Washington, DC: Transportation Research Board, The National Academies of Sciences, Engineering, and Medicine, available at http://onlinepubs.trb.org/onlinepubs/tcrp/tcrp_webdoc_6-b.pdf, accessed 29 July 2017.

UIC/CER (2015) *Rail transport and Environment: Facts and Figures*, Paris: UIC – ETF (Railway Technical Publications), September.

US Department of Energy (n.d. a) Gasoline Vehicles - Learn More About the New Label, available at https://www.fueleconomy.gov/feg/label/learn-more-gasoline-label.shtml accessed 10 April 2017.

US Department of Energy (n.d. b) 2015 Porsche Cayman GTS, available at https://www.fueleconomy.gov/feg/noframes/35151.shtml, accessed 13 June, 2017.

US Department of Energy (2015) *Average fuel economy of major vehicle categories*, Washington, DC: US Department of Energy, available at https://www.afdc.energy.gov/data/10310, accessed 30 July 2017.

US Department of Energy (2017a) *2017 Best and Worst Fuel Economy Vehicles*, available at https://www.fueleconomy.gov/feg/best-worst.shtml, accessed 11 August 2017.

US Department of Energy (2017b) *New All-Electric Vehicles*, available at https://www.fueleconomy.gov/feg/PowerSearch.do?action=noform&path=3&y earl=2017&year2=2018&vtype=Electric&srchtyp=newAfv&pageno=2&sortBy=Comb&tabView=0&row Limit=10, accessed 11 August 2017.

US Department of Energy (2017c) *Electric Vehicles Learn More About the New Label*, available at https://www.fueleconomy.gov/feg/label/learn-more-electric-label.shtml, accessed 11 August 2017.

US Department of Energy (2017d) *Emissions from Hybrid and Plug in Electric Vehicles*, Alternative Fuels Data Center, 18 May, available at https://www.afdc.energy.gov/vehicles/electric_emissions.php, Accessed 14 August 2017.

Vale R and Vale B (2009) *Time to Eat the Dog? The Real Guide to Sustainable Living*, London: Thames and Hudson.

Vale R and Vale B (eds) (2013) *Living within a Fair Share Ecological Footprint*, London: Earthscan.

VW (2013) *XL1*, Wolfsburg: Volkswagen Produktkommunikation.
WCC (n.d.) *Greenhouse gas emission reduction targets*, available at http://wellington.govt.nz/services/environment-and-waste/environment/climate-change/greenhouse-gas-emission-reduction-targets, accessed 8 July 2017.
Yang Z, Zhu L and Bandivadekar A (2015) *A Review and Evaluation of Vehicle Fuel Efficiency Labeling and Consumer Information Programs*, Singapore: Asia Pacific Economic Cooperation Secretariat.

12
ENERGY

Brenda Vale, Robert Vale and Fabricio Chicca

Sources of energy

Without the energy of the sun, there would be no life on earth. For many centuries, human society depended entirely on the sun's energy, so it should be no surprise that the sun was worshipped as a god in societies as far apart as Ancient Greece and the Mayan civilisation in South America. Malanima (2014:7) put it simply: "for 85–90% of human history, food was the only source of energy." The advent of fire 1 million to 500,000 years ago (Malanima, 2014: 7) meant that wood—stored solar energy—could be burnt for cooking, lighting and heating. Water wheels (used for at least the last 2,000 years Rao, 2011: 9]) and wind mills, which first appeared circa the seventh century CE from Asia (Malanima, 2013: 67), introduced new sources of energy. However, the flow of water is still driven by the sun through the hydrological cycle, and the winds also depend on the sun heating the air and the turning of the earth. These new sources of energy only increased energy availability by 1–2 per cent (Malanima, 2014: 9). Society still depended on the sun and its by-product—wood.

The growth and decline of cities has been linked to the growth and decline of civilisations (Atkinson, 2014), and this urbanisation initially also relied on the energy of the sun.

"Roughly 10 percent of England's population were living in towns by the year 1000, which meant the country's farming methods had developed the efficiency to produce a 10 percent surplus—while the town dwellers were generating sufficient profit to purchase the foodstuffs and other supplies they needed" (Lacey and Danziger, 1999: 87).

The fuel for this society was wood, which could be turned into charcoal to produce the high temperatures needed to work metal. Wood in the past was also the material for building and for transport in the form of wooden carts with wooden wheels. The other energy entering this way of living also came from the sun, in the food eaten by people and animals doing the required work of farming, building and other trades, although the most important was farming. Without food, people starved, and this happened, for example, when an eruption in Iceland blocked the sun for several years around the year 1000 (Sothers, 1998). Without a reliable source of energy, in this case the sun, human society struggles.

Energy still pervades everything we do, though wood now forms only a relatively small proportion of global energy. In 2014, biofuels (including wood) and waste accounted for 10.3 per cent of global primary energy, compared with coal, 28.6 per cent; oil, 31.3 per cent; natural gas, 21.2 per cent; nuclear, 4.8 per cent; hydro, 2.4 per cent; and other (including geothermal, solar and wind), 1.4 per cent (IEA, 2016: 6). Alvarez and Smolker (2014) suggest approximately two-thirds of biofuel use is wood and other biomass (for example, dried cow dung) used for heating and cooking. Humanity has moved from a sustainable if somewhat precarious reliance on energy derived from the sun to a heavy reliance on the "fossil fuels" of coal, oil and natural gas.

Fossil fuels and nuclear energy

Fossil fuels are perhaps best viewed as super-concentrated solar energy, which makes them so very useful, as a small quantity can yield a lot of power. They are ancient plants and animals (the fossils) that have decayed and been compressed in layers in the surface of the earth (Black, 2012: 6). Because of the concentrated nature of fossil fuels, the Canadian energy analyst Dave Hughes worked out that it would take about 8.6 years of normal working hours for a person on a bicycle to generate the energy contained in a barrel of crude oil (6 gigajoules or 1,700 kilowatt hours) (The Green Interview, 2010).

This same barrel of oil contains the energy of 161 cubic metres of natural gas (National Energy Board Canada, 2016). To give a sense of what this means, Table 12.1 sets out what you could do with 6 gigajoules of energy in some of the places in the stories.

Table 12.1 shows how concentrated fossil fuels are. To run the sauna for a 15-minute session, including the hour to heat it up, would mean that the cyclist would have to pedal for 2.2 months. In contrast, to run a DVD player with a bicycle is possible. A normal person can generate 100 watts when cycling (Bluejay, 2016), while a DVD player uses only 30 watts (Efficiency Vermont, 2016).

Coal

Coal is probably the oldest fossil fuel to be exploited. It was mentioned in 371 BCE, and coal cinders have been found in the ruins of Roman settlements. In 1243, a charter was granted to mine coal in Newcastle in the United Kingdom for industrial use (Wright, 1964: 62). Use of coal in the United Kingdom was spurred on by declining supplies of timber, with wood shipped in from Scandinavia as early

TABLE 12.1 What you can do with the energy in a barrel of oil

Place	Activity	For how long	Reference
United States	Run chest freezer	3.3 years	Efficiency Vermont (2016)
Germany	Run sauna (traditional type)	443 sessions*	Finnleo (2016)
Brazil (rural)	Using pick-up truck	22 hours (1,321 km)**	Packer (2011) Consumo Combustível (2013)
New Zealand	Using quad bike	131 hours (2,622 km)***	Rickard (2012) Worksafe NZ (2017)
Urban Japan	Heating mattress daily (kotatsu)	11.6 years	400 watts for 1 hour/day
United Kingdom	Gas water heating (combi boiler)	3 months	Office of Energy Efficiency and Renewable Energy (n.d.a)
South Sudan	Run DVD player	94 years	Efficiency Vermont (2016)

*1 hour to heat up for a 15-minute sauna
**1,700 kilowatt hours is how much fuel?
1 litre of petrol is 9.4 kilowatt hours
1 litre of diesel is 11.1 kilowatt hours (Packer, 2011)
    So 1,700 kilowatt hours is 180.9 litres of petrol or 153.2 litres of diesel.
    A Ford F-250 XLT 3.9 Turbo 4×4 CS 2010 Diesel has figures of 6.7 kilometres/litre urban and 8.8 kilometres/litre rural (Consumo Combustível, 2013) which works out at 14.9 litres/100 kilometres urban and 11.6 litres/100 kilometres rural. The rural figure will be used here. So the 1,700 kilowatt hours will drive the Ford ute (or pickup truck) 1,321 kilometres. Assuming an average speed of 60 kilometres/hour gives 22 hours of driving.
*** In a test (Rickard, 2012), the following fuel consumption figures were achieved by a selection of quad bikes.

ATV fuel economy over a 5 kilometre tarmac course

	Polaris	Suzuki	Yamaha	Honda	Kawasaki
Bike only	480 millilitres	400 millilitres	300 millilitres	320 millilitres	230 millilitres

This is an average of 346 millilitres for 5 kilometres, or 6.9 litres per 100 kilometres. This means that an average petrol quad bike could go 2,622 kilometres on the energy in a barrel of oil, approximately twice as far as the pickup.

as 1230. In the thirteenth century, both pit coal (mined) and sea coal (collected from the beaches of Northumberland and Durham, where the coal seams were exposed) were used (Gimpel, 1976: 80–81). The expanding population of London in the sixteenth century and the shortage of fuel wood meant that sea coal, shipped down the coast from northeast England, had to be used not just for industrial purposes but also for domestic heating. Because of its smell, "Well-bred ladies would not even enter rooms where coal had been burnt, let alone eat meat that had been roasted over a sea-coal fire" (Brimblecombe, 1987: 30). This increasing use of coal led to increasing complaints about air pollution (Mosley, 2014: 147), so from the start, there were problems with the use of fossil fuels.

    The big expansion in the use of coal came in the early nineteenth century and fuelled a dramatic rise in energy consumption in Western Europe from around

1,600 petajoules to over 12,200 petajoules (Warde, 2014: 131–132). This led to both a population expansion in Western Europe, with the population of England increasing the most from just over 9 million in 1800 to over 23 million by 1900 (Grigg, 1980:61), and to an increase in energy use from around 22 gigajoules/person to 65 gigajoules/person (Warde, 2014: 134). In 2001, the primary energy figure for Western Europe was 160 gigajoules/person (Giere and Stille, 2004), showing that we are still energy-hungry, despite improvements in the efficiency of energy conversion. This nineteenth-century expansion in energy use depended on coal, which could not have been extracted without Newcomen's coal-fired steam engine to pump out the coal mines. Newcomen's engine appeared in 1705, and between 1700 and 1800, coal production increased by a factor of five. Watt, in 1781, improved the steam engine, cutting its consumption of coal by two-thirds (Dahlen, 2010). Such improvements in conversion efficiency only highlight the significant rise in overall energy consumption in the eighteenth century.

Coal is still important, providing 37 per cent of the world's electricity generation. It is seen as a secure energy source, widely available from politically stable suppliers (IEA, 2000: 5). Recoverable coal reserves are known in at least 70 countries (World Coal Association, 2017). Despite this, coal is the only fossil fuel to experience a slight dip in consumption between 1971 and 2015 (IEA, 2016: 14). In 2015, the largest producer of coal was the People's Republic of China, at 45.8 per cent of the global total, followed by the United States at 10.6 per cent. At the current consumption rate, the estimated 1.1 trillion tonnes of known reserves would last 150 years at the current rate of coal production, in contrast to oil and gas reserves, which could last 50 and 52 years, respectively, at current production rates (World Coal Association, 2017). The problem is whether extracting and using coal at this rate is compatible with holding the global temperature rise to 2°C or less.

As a carbon-based fuel, burning coal contributes to greenhouse gas (GHG) production. For every tonne of coal burned, 2,419 kilograms of CO_2 are produced (Davies, n.d.). As each tonne of coal produces on average 29,208 megajoules of energy, each megajoule of coal is associated with 82 grams of CO_2. However, methane, another GHG, is released during underground mining. Burnham et al. (2012) estimate that each megajoule of coal has additional extraction emissions of 0.138 gram of methane. Globally, in 2015, GHG emissions from burning coal produced 46 per cent of all CO_2 emissions, 31 per cent of which came from electricity generation plants. The top four countries for emissions from coal were China, the United States, the European Union and India (Olivier et al., 2016: 5). This underlines the importance of coal, even if in 2015 there was a 1.8 per cent decrease in its use (Olivier et al., 2016: 4).

Oil

The wide exploitation of oil has a much shorter history than coal. Nevertheless, from ancient times, where "crude oil" appeared near the surface, it was used. The Mesopotamians made use of oil seepage (Black, 2012: 21), and Native Americans used the mastic properties of asphalt from surface deposits in Kentucky (Collins, 1981).

Drilling for oil seems to have commenced in the middle of the nineteenth century in both Europe and North America (Zabawski, 2011). This oil was used in the making of kerosene, formerly made from coal, that could be burned for lighting (Black, 2012: 26, 31). Kerosene was a substitute for whale oil, which had taken over from animal fats for lighting. However, it was the use of oil for moving things that led to the huge expansion of the oil industry in the twentieth century, and this expansion was dominated by the United States (Painter, 2012).

In 1900, world oil production was 400,000 barrels/day, while at the end of the twentieth century, this had reached 74 million barrels/day (Hughes and Rudolph, 2011). Until the 1970s and the oil wars, this growth followed an exponential pattern, approximately doubling every 10 years. From 2015 to 2016, world oil production was steady at 93.7 million barrels/day (IEA, 2017a: 3). In 2015, more oil was used than any other fuel, forming 31.8 per cent of world energy (IEA, 2017a: 7). The reason for the continued popularity of oil is that it is so concentrated, it can be used to power vehicles and mobile machinery. Oil is currently used for electricity, for heat generation and to power transportation. Renewables can be used for the first two applications, but as a substitute for the third, renewables are much more problematic (Reynolds, 2014). The world energy balance shows that road transport is 93.7 per cent oil-powered, 3.7 per cent biofuels, 2.0 per cent natural gas and 0.5 per cent electricity (IEA, 2017a: 8).

There is also no shortage of oil. As mentioned above, at current usage, there are supplies to last at least another 50 years. Like coal, significant oil reserves are found on every continent. In 2014, liquid and solid fuels (which include oil and coal) accounted for 75.1 per cent of global GHG emissions, and these came from burning the fuels and from using them for cement manufacture (Boden et al., 2017).

Because of the usefulness of petroleum for moving vehicles, oil is also extracted from coal and from deposits that are harder to exploit, such as oil shales. This has led to the investigation of the energy return on investment, normally shortened to EROI, defined as "the net energy produced during the life of a system divided by the total energy input to build and run the system" (Kreith, 2012). This obviously applies to all energy sources. There is a debate over whether oil shales are worth using because of the energy required to extract oil products from them, leading to an EROI of around 1.5:1. O'Connor and Cleveland (2011) state: "Oil shale unambiguously emits more greenhouse gases than conventional liquid fuels from crude oil feedstocks by a factor of 1.2 to 1.75". Earlier, Brandt (2008) estimated that GHG emissions from petroleum products from oil shales would be 21–47 per cent higher than those from conventional sources, suggesting that exploiting such unconventional oil reserves is not going to be a solution to keeping global temperature rise to 2°C or less, although it may be good for making money.

Natural gas

Natural gas has the shortest history of exploitation of the three fossil fuels, although the Chinese drilled for natural gas around 600 BCE (Robinson, 2006: 2), piped it

TABLE 12.2 Emissions from direct burning of selected fossil fuels (Quaschning, 2015)

Fuel	Kilograms CO_2/gigajoule
Hard coal	94.6
Diesel	74.1
Crude oil	73.3
Petrol (gasoline)	69.3
LPG	63.1
Natural gas	56.1

using bamboo and used it to boil seawater for the extraction of salt. The oracle at Delphi has also been linked to lightning setting fire to seepage of natural gas, producing a tongue of flame (NaturalGas, 2013). The nineteenth-century spread in the use of gas in cities was coal gas, originally a by-product of the process used to turn coal into smokeless coke. Burning coal gas in a lamp gave light, although heat was a by-product (Shepherd and Shepherd, 2014: 201). With the advent of electric lighting, coal gas was used for cooking and heating (Barty-King, 1984). In the 1960s in the United Kingdom, the move was from gas made from coal to supplies of natural gas from under the North Sea. In terms of UK primary energy, North Sea gas rose from a 5 per cent share in 1970 to a maximum of 43 per cent in 2010 (DBEIS, 2017: 9).

By 2014, natural gas accounted for 21.2 per cent of the world energy supply (IEA, 2016: 6) and 18.5 per cent of all emissions from fossil fuels (Boden et al., 2017). Table 12.2 compares emissions from fossil fuel sources (Quaschning, 2015) to show the relative advantage of natural gas.

Nuclear energy

Of all the conventional energy technologies, nuclear is perhaps the most contentious. In 2014, nuclear plants produced only 4.8 per cent of global energy, up from 0.8 per cent in 1974 (IEA, 2016: 6). In the same year, the Intergovernmental Panel on Climate Change (IPCC) (2014: 526) stated that there were 130 years of uranium at this rate of use, and that this could be extended to 250 years if all uranium was included. These limitations and the small contribution made by nuclear energy suggest that even if there could be a rapid expansion in its use, it is not going to replace fossil fuels. Nuclear energy plants generate electricity, and apart from submarines and ships, nuclear-powered engines have not made their way into vehicles—except in the movies (Paramount, 2012).

In 2014, the OECD countries were responsible for 78.1 per cent of all nuclear energy, with the United States, France and Japan having the highest installed capacities at 99 gigawatts, 63 gigawatts and 42 gigawatts, respectively (IEA, 2016: 16–17). This was 19.5 per cent of US electricity generation in the same year (EIA, 2017), 78.4 per cent of French electricity (IEA, 2016: 17) and just under 1% of Japan's, although in 2011, nuclear energy had formed nearly 30 per cent of Japan's electricity and there were plans for an increase to 40 per cent by 2017 (World Nuclear Association, 2016). This dramatic drop was due to the Fukushima accident, which led to "wide public protests calling for nuclear power to be abandoned" (World Nuclear Association, 2016). The continuous production of GHG is equally endangering human survival, but GHG does not provoke the same perceived threat as a nuclear accident. Both GHG and nuclear radiation are equally invisible, and it is perhaps the lingering images of the damage nuclear weapons can do that produces this paradox. Despite the promises of the 1950s, when nuclear power appeared to be the future of energy supply, with President Eisenhower's "Atoms for Peace" (Beaver, 2011), nuclear power as we currently know it will never be able to be the cornerstone of world energy supplies. Despite this, fusion power (current nuclear power generation uses fission technology) is still seen as the future of human energy supply by some (Highfield et al., 2009), as the alternative renewable energy technologies seem neither as concentrated nor as convenient as fossil fuels.

Renewable energy sources

In terms of world energy, in 2015, biomass fuels and waste formed 15.8 per cent of total energy, hydro 2.2 per cent and solar, wind and geothermal part of the 2.1 per cent "other" category (IEA, 2016: 7). If the world is really going to rely on renewables, there is a very long way to go.

Biomass

Biomass means fuel that is grown, from wood to ethanol made from sugar cane. The only relatively concentrated and renewable source of energy is the source that underpinned human development, wood. Moves have been made in the developed world to replace fossil fuels with wood. Erakhrumen (2011) states that wood accounts "for more than 16 percent of total energy supply in Sweden and Finland, and 12 to 18 percent in some Central and East European countries".

Wood has been used as a fuel for trains, but even with abundant wood from the vast forests of the United States, the more concentrated nature of coal was appreciated for powering railroad locomotives as "2000 pounds [917kg] of coal equalled 5,250 [2381kg] pounds of wood" (White, 1979: 83–90). Such direct burning is not the only way that wood has been used to power vehicles. In the 1920s, wood gas generators were invented and during World War II, when there was a shortage of petrol, many of these were used to power vehicles. Sweden ran 40 per cent of its vehicular fleet on wood gas during this period (Shrinivasa and Mukunda, 1984).

However, it is fair to say that most of the world's fuel wood is burned for heat, and especially to provide heat for cooking. Behera et al. (2015) report that "globally about 2.7 billion people still rely on traditional biomass as their main source of energy for cooking and heating". With an increase in income, the move is away from biomass fuels to LPG, while in parts of the developed world, the quest is to burn more wood for heat and electricity. Burning wood is considered carbon neutral in Europe, but this has led to harvesting wood from US forests, turning this wood into pellets and then shipping these to Europe (Upton, 2015), with a resulting increase in GHG emissions. Wood is only carbon neutral if harvested cyclically, such that the trees absorb CO_2 at the same rate at which it is released by burning.

Hydro power

Hydro power has the longest history among the renewable sources of electricity and accounted for 2.2 per cent of global primary energy in 2015. The first water-powered light bulb was lit at Cragside, Northumberland in 1878 (Vale and Vale, 2005). In 1882 in Wisconsin, the first hydroelectric power plant to serve private and commercial customers was opened (IHA, 2016). Even New Zealand was early to use hydro, with the first street lighting in the small town of Reefton being powered by hydro in 1888 (Reefton Tourism and Cloake Creative, 2017). In 2015, hydro supplied 9.6 per cent of primary energy in New Zealand (geothermal supplied 22 per cent) (MBIE, 2015: 47), and 57 per cent of the electricity generated (MBIE, 2015: 53). In 2014, Norway was the country with the greatest hydro component in electricity (86 per cent), followed by Venezuela (68.3 per cent), Brazil (63.2 per cent) and Canada (58.3 per cent). In the same year, The People's Republic of China generated the most hydro at 26.7 per cent of the world total, and this formed 18.7 per cent of their electricity (IEA, 2016: 19).

The flooding of land means that hydro has an environmental impact through the methane that is released when the land is initially submerged and the plants that cover it rot under the water (Kumar and Sharma, 2017). Overall levels of emissions from hydroelectricity are still an area of debate, especially given that only 17 per cent of the world's potential hydropower sites have been utilised (Arntzen et al., 2013: 1).

Solar

The use of solar energy to warm buildings goes back to the Greeks and Romans. The modern idea of heating houses using the sun through appropriate design and orientation goes back to the Keck brothers in the United States (Butti and Perlin, 1980: 180–195). It is easy to heat a house when the sun is shining, the problem being how to store this energy for periods when the sun is covered by cloud or at night. Heat storage has been proposed in tanks of water, phase change chemicals, rock beds and insulated structural mass in the form of concrete or masonry (Vale and Vale, 1975: 29–40), with the latter now the most common method used, not least because it is the cheapest.

Solar energy has also been used to heat water directly and this also has a history stretching back to the end of the nineteenth century in the United States (Butti and Perlin, 1980: 114–127). In 2013, China had by far the largest share of global solar water heating at 70 per cent of the total, followed by the United States (4.5 per cent), Germany (3.3 per cent), Turkey (2.9 per cent) and Brazil (1.8 per cent)

(Worldatlas, 2017). In 2014, the entire world's solar water heaters generated approximately 414 gigawatts thermal (Worldatlas, 2017). Although solar water heaters are simple and only generate heat, they form an important, if somewhat forgotten, renewable resource.

Like solar water heaters, photovoltaic technology (PV) was developed in the late nineteenth century but it was the space race of the 1960s that spurred modern development (Mir-Artigues, 2016: 37–42). Its installation on rooftops and other surfaces has been exponential in recent years, rising from around 15 gigawatts installed capacity in 2008 to approximately 320 gigawatts in 2016. This is distributed with 6 per cent in Italy; 13 per cent each in Germany, Japan and the rest of the world; 14 per cent in the rest of Europe; 15 per cent North America, and 26 per cent in China and Taiwan (Fraunhofer Institute, 2017). Most PV panels are made in China, and because of low-cost government loans for expanding this industry, prices have dropped globally (Fehrenbacher, 2015), falling by 80 per cent (Fialka, 2016). However, because of China's current dependence on fossil fuels, Yao et al. (2014) estimated that emissions from PV panels made in China could be as high as 207 grams CO_2/kilowatt hour, although Yue et al. (2014) suggest a much lower figure of 72 grams CO_2/kilowatt hour, only slightly higher than values from similar studies for the United States and Europe.

Wind

The generation of electricity from wind goes back to early experiments in the 1880s in Scotland and the United States, after which Denmark took the lead (Nixon, 2008). The first large-scale wind generator was 1.25 megawatts and was erected in Vermont in 1941; this was the first machine to feed electricity directly into the local grid (Renewable Energy Vermont, 2017). The largest modern machines are rated at approximately 3 megawatts. China again leads the world in terms of installed wind capacity. In 2016, this was 169 megawatts, with the United States at 82 megawatts, Germany 50 megawatts, India 28 megawatts, Spain 23 megawatts and the United Kingdom 15 megawatts. Total world installed capacity in 2016 was 487 megawatts (WWEA, 2017).

Geothermal

Naturally occurring hot springs have been used by people for centuries, and in 1892, water from hot springs was used in Boise, Idaho, to heat 200 homes and 40 commercial premises (Office of Energy Efficiency and Renewable Energy, n.d.b). The first plant to generate geothermal electricity was the 1911 plant at Lardello in Tuscany, Italy, which is now the world's second largest geothermal power plant (Kable, 2017). More than 90 per cent of the world's 10 gigawatts of geothermal capacity comes from countries that are young geologically and with active volcanic fields (Bayer et al., 2013). Bertani and Thain (2002: 2–3) reported emissions of 4–740 grams CO_2/kilowatt hours across a large number of plants operating in

2001. A similar range for New Zealand alone was found by Rule et al. (2009), with an average of 80 grams CO_2/kilowatt hour and a range of 30–570 grams CO_2/kilowatt hour, depending on the depth of the well, again showing that even renewable sources of energy have environmental impacts.

Electricity

Electricity is the one type of energy that appears in all the stories except one. Lifecycle GHG emissions from the generation of electricity from various sources are compared in Table 12.3, which also looks at each source's share of global electricity generation in 2015 (IEA, 2017b: 3).

What Table 12.3 emphasises is that that there is no easy answer to how electricity can be generated to satisfy current rates of consumption without having some environmental impact. It is hard to see how impact can be regulated in order to encourage the most appropriate or lowest impact sources of power generation. Left to a free market where profit is the aim, the consumer has no idea what impact their consumption of electricity may be having. Even buying "green power" could mean a difference in impact of almost a hundred times (going from the best wind in Table 12.3 to the worst geothermal), although Table 12.3 shows that, on the whole, electricity from renewables and nuclear is much lower in impact than electricity from fossil sources. Since the values in the table are from the World Nuclear Association, perhaps this is not surprising.

TABLE 12.3 Comparison of emissions from generation of electricity

Source of electricity	Per cent breakdown of electricity 2015	Tonnes CO_2/ gigawatt hours low★	Tonnes CO_2/ gigawatt hours average★	Tonnes CO_2/ gigawatt hours high★
Coal	39.3	756	888	1,310
Natural gas	22.9	362	499	891
Oil	4.1	547	733	935
Hydro	16.0	2	26	237
Nuclear	10.6	2	29	130
Biofuels and waste	2.2	10	45	101
Wind/solar/ geothermal/tidal	4.9			
Wind		6	26	124
Solar PV		13	85	731
Geothermal★★		30	80	570

★World Nuclear Association (2011)
★★Rule et al. (2009)

Domestic energy in the stories

Table 12.4 sets out the sources of domestic energy from the stories. The amount of energy used will vary according to the climate; the aim here is to look at the degree to which people in the stories rely on fossil fuels.

TABLE 12.4 Sources of energy in the stories and energy EF in global hectares

Place	Gas	Oil	Coal	Electricity	Wood/biomass	Renewables (wind/solar)	EF global hectares/person
Urban Argentina (1)	√			√			0.38
Urban Australia				√			0.42
Rural Brazil				√			0.23
Urban Brazil	√			√			0.71
Urban Canada				√			0.24
Urban Cuba				√			0.08
Urban Finland				√	√		0.58
Village Germany		√		√		√	2.96
Rural India		√ (kerosene)		√			0.20
Rural Indonesia					√		0.01
Urban Japan		√		√			0.64
Urban Malaysia	√			√			0.39
Urban Mongolia			√	√			0.40
Urban Morocco				√			0.12
Rural Mozambique				√			0.02
Rural Myanmar				√			0.02
Rural New Zealand				√	√		3.93
Urban South Africa				√			0.18
Urban South Sudan				√			0.03
South Sudan UNMISS				√			0.04
Urban Tonga				√		√	0.20
Urban United Kingdom	√			√			1.77
Rural United States				√			0.90
Urban United States				√			0.13

Apart from the Baduy in rural Indonesia, who avoid the attributes of modern life, the dependence on electricity in all other stories is perhaps a surprise. Only village Germany and urban Tonga use solar water heaters. Wood as a fuel features in three of the stories, two rural and one urban (Oulu, Finland, where it is used to heat the sauna).

Energy and land

In the end, renewable energy comes back to land. This is explored in Table 12.5. In south Germany, it is common to see PV arrays in fields beside the railway. However, The Fraunhofer Institute (Wirth, 2017: 39) states that PVs are only installed on brown-field sites or on land adjacent to roads and railway lines, and that there is no conflict for land. That said, food crops are still grown next to railway lines and roads in many areas.

The big output comes from wind or, to a lesser extent, PV panels in southern Germany, whereas ethanol at an average 85 gigajoules/hectare and forest at 75.2 gigajoules/hectare are of the same, much lower, order of magnitude. Wind turbines can harvest energy in three dimensions and this gives them an advantage, but it also makes them very visible, which may explain the rush to install PVs in fields. The disadvantage of PVs, as mentioned above, is that the electricity is not always generated when it is needed, whereas both wood and ethanol are fuels that can be stored until needed. There is no easy answer to the convenience of fossil fuels, but although you cannot live without food, you can, as the low-impact stories demonstrate, live with very little energy and without electricity.

Carbon footprint

To extract the true carbon footprint (CF) from the stories is not straightforward as energy tends to be involved in most things we do. If you walk everywhere and only eat plant-based crops, which will give you a very low EF, you are using the energy of the sun to support your lifestyle. However, you might still have to account for the non-solar energy bound up in the materials of your house, or the clothes you wear. If you build a mud house and roof it with iron sheets from a demolished building and only wear second hand clothes, then you could argue that the fossil fuel energy used to produce these belong to the person who bought the materials in the first place and discarded them. In this way it might be possible to achieve a zero carbon EF since you are not producing any GHG emissions. As soon as you start buying consumer goods you have a CF. If you also start eating dairy products there is a GHG element in your food that comes from the methane emitted by ruminants and this will add to your CF. Add in a house of reinforced concrete, with heating or cooling using electricity from at least some non-renewable sources and travel to work in a steel and plastic car and your CF quickly starts to grow.

If you live in a high-income country you are much more likely to have a high CF. In 2012 in these countries the CF formed approximately two-thirds of the

TABLE 12.5 Energy that can be generated from a hectare in southern Germany

Crop	Annual yield	Reference	Annual yield	Reference
PV panels (tracking system)	1,190 kilowatt hours/kilowatt-peak 1 megawatt-peak/hectare 1,190 megawatt hours/year	Fraunhofer-Institute for Solar Energy Systems (2015:53) Narasimham (2014)	4,284 gigajoules	
Forest	9.5 cubic metres (annual increment)	Federal Ministry of Food, Forestry and Consumer Protection (2011)	19.8 megajoules/kilogram 75 gigajoules	Ashton and Cassidy (2007: 189–192) Assuming density 400 kilograms/cubic metre
Wind plus ethanol from sugar beet (see below)*	1.7 megawatts/hectare for wind farm A 1-megawatt machine on land generates 2,000 megawatts annually	Vale and Vale (2009: 93–94) EWEA (2012)	7,200 gigajoules	
Ethanol from sugar beet	2,000–3,000 litres gasoline equivalent	IEA (2007)	85 gigajoules	Assuming 20 megajoules/kilogram

*Based on the assumption that the land of a wind farm can be used to grow crops

EF, while for middle income countries it is just over half and perhaps a fifth in low income countries (WWF, 2016: 80). For example, the average Australian emits 20.6 tonnes CO_2/capita, partly because of the coal used to generate electricity, while for a person in the UK the average emission is 9.7 tonnes/capita, as more electricity is made from natural gas. It drops to 1.2 tonnes/capita in India and to only 0.3 tonnes/capita in Kenya (Guardian News, 2016). The value of the CF is it shows clearly the contribution each individual makes to climate change (Galli et al, 2011: 22).

Table 12.6 sets out the CFs of the stories. This has been achieved by adding up the GHG implications (CO_2 and CH_4 only) of the ecological footprints (EFs) (food, energy, travel, housing, consumer goods). This is not the full CF, as it lacks the contribution of other GHGs such as nitrous oxide and ozone or the CF of government services and infrastructure, which in turn depend upon the country where each story is set. However, it is useful to compare it to the EF. Average 2006 per capita CO_2 emissions for each country are given for comparison (Guardian News, 2016), and the CFs in the table are set out from largest to smallest.

Three stories have a CF bigger than the national average, accepting that these 2006 average figures are old. The family in urban São Paulo has a marginally larger

TABLE 12.6 Comparison of carbon and ecological footprints per person

Place	EF global hectares	CF tonnes GHG/annum	2006 average tonnes CO_2
United States rural	4.30	10.44	19.8
Canada	4.03	6.76	18.8
Australia	3.84	5.76	20.6
Brazil rural	5.90	5.45	2.0
United Kingdom	7.61	4.93	9.7
Mongolia	4.94	4.93	2.9
Germany	10.47	3.88	10.4
New Zealand	8.84	3.44	9.4
Argentina	3.43	3.04	4.1
United States urban	4.11	2.81	19.8
Finland	3.81	2.33	11.2
Malaysia	4.38	2.32	6.7
South Africa	1.64	2.31	10.0
Brazil urban	2.31	2.30	2.0
Tonga	1.71	1.57	N/A
Morocco	2.66	1.51	1.0
Cuba	0.83	1.29	2.5
Myanmar	1.09	1.22	0.3
Japan	2.06	1.09	9.8
South Sudan UNMISS	0.16	0.24	0.3
India	0.33	0.19	1.2
South Sudan (Juba)	0.10	0.12	0.3
Indonesia	0.02	0.08	1.2
Mozambique	0.13	0.06	0.2

CF (2.30 tonnes/person as opposed to 2.00 tonnes) while the rural farming family in Brazil are 5.45 tonnes/person. As the CF here does not include the full CF of living in Brazil, many people in Brazil must be living a very low-energy lifestyle. The family in Ulaanbaatur have a CF of 4.93 tonnes/person as opposed to the 2006 average of 2.9 tonnes. This may reflect the increased CF from living in an urban area and the rising living standard that comes with this. The family from Japan have a CF of 1.09 tonnes/person as opposed to a 2006 average for Japan of 9.8 tonnes. The family live a modest life in terms of transport and overall use of domestic energy, though their average is helped by being a family of two adults and three young children. This may also be true of Malaysia, with their four children, as their CF is 2.32 tonnes against a 2006 national average of 6.7 tonnes, and South Africa, with three children (2.31 tonnes/person compared with 10.0 tonnes/person).

Comparing EF and CF, rural United States and rural Germany emerge as two anomalies. The German story, with its very high EF and relatively low CF, is not so much the lifestyle as being German. As explained in Chapter 8, the multiplier for Germany is high because many cities have been built on the most fertile land. This sends a warning regarding global urbanisation. Building on productive land will increase the EF because of the loss of the ability to absorb CO_2. Rural United States has the opposite pattern of a high CF and lower EF, again to do with the ability of the nation's land to sequester CO_2.

As the low impact stories show, the real way to reduce CF turns out to be not through innovation and improving energy efficiency and effectiveness, but in not using much energy beyond that obtained from food. As Chapter 11 reveals, the lowest-impact form of transport is the bicycle (and the electric bicycle). When it comes to domestic energy use, then, sharing is perhaps the key. There is a big difference between having five TVs in a house, as in the urban United States story, and watching your parents' TV through their window while sitting outside in the street, as in the story from Havana. Sharing is social and the majority of people are social animals.

References

Alvarez I and Smolker R (2014) A global overview of wood based bioenergy: Production, consumption, trends and impacts, Global Forest Coalition, available at http://globalforestcoalition.org/wp-content/uploads/2010/06/REPORT-WOOD-BASED-BIOENERGY-FINAL.pdf, accessed 11 August 2017.

Arntzen E V, Miller B L, Niehus S, O'Toole A C and Richmond M (2013) *Evaluating greenhouse gas emissions from hydropower complexes on large rivers in Eastern Washington*, Pacific Northwest Laboratory, available at http://www.pnl.gov/main/publications/external/technical_reports/PNNL-22297.pdf, accessed 18 August 2017.

Ashton S and Cassidy P (2007) *Energy Basics*, in Hubbard W, Biles L, Mayfield C and Ashton S (eds) *Sustainable Forestry for Bioenergy and Bio-based Products: Trainers Curriculum Notebook*, Athens, GA: Southern Forest Research Partnership, Inc, pp. 189–192.

Atkinson A (2014) Urbanisation: A brief episode in history, *City*, 18(6), pp. 609–632.

Barty-King H (1984) *New Flame: How Gas Changed the Commercial, Domestic and Industrial Life of Britain Between 1813 and 1984*, Tavistock, UK: Graphmitre Ltd.

Bayer P, Rybach L, Blum P and Brauchler R (2013) Review on life cycle environmental effects of geothermal power generation, *Renewable and Sustainable Energy Reviews*, 26, pp. 446–463.

Beaver W (2011) The failed promise of nuclear power, *The Independent Review*, 15(3), pp. 399–411.

Behera B, Rahut D B, Jeetendra, A and Akhter A (2015) Household collection and use of biomass energy source in South Asia, *Energy*, 85, pp. 468–480.

Bertani R and Thain I (2002) Geothermal Power Generating Plant CO_2 Emission Survey, *International Geothermal Association (IGA) News*, available at file:///C:/Users/Brenda/Downloads/iganews_49a.pdf, accessed 21 August 2017.

Black B C (2012) *Crude Reality: Petroleum in World History*, Lanham, Maryland: Rowman and Littlefield Publishers, Inc.

Bluejay M (2016) Generating electricity with a bicycle, available at http://michaelbluejay.com/electricity/bicyclepower.html, accessed 14 August 2017.

Boden T A, Marland G amd Andres R J (2017) *Global, regional, and national fossil fuel CO_2 emissions*, available at http://cdiac.ornl.gov/trends/emis/tre_glob_2014.html, accessed 17 August 2017.

Brandt A R (2008) Converting oil shale to liquid fuels: Energy inputs and greenhouse gas emissions of the Shell in situ conversion process, *Environmental Science and Technology*, 42, pp. 7489–7495.

Brimblecombe P (1987; 2011 edition) *The Big Smoke*, London: Routledge.

British Petroleum (BP) (2017) Solar energy, available at http://www.bp.com/en/global/corporate/energy-economics/statistical-review-of-world-energy/renewable-energy/solar-energy.html, accessed 20 August 2017.

Burnham A, Han J, Clark C, Wang M, Dunn J and Palou-Rivera I (2012) Life-cycle greenhouse has emissions of shale gas, natural gas, coal, and petroleum, *Environmental Science and Technology*, 46(2), pp. 619–627.

Butti K and Perlin J (1980) *A Golden Thread: 2500 Years of Solar Architecture and Technology*, London: Marion Boyars.

Collins M B (1981) The use of petroleum by late archaic and early woodland peoples in Jefferson County, Kentucky, *Journal of Field Archaeology*, 8(1), pp. 55–64.

Consumo Combustível (2013) Consumo F-250 XLT 3.9 Turbo 4×4 CS 2010, available at http://consumocombustivel.com.br/consumo-f-250-xlt-3-9-turbo-4x4-cs-2010/, accessed 12 June 2017.

Dahlen M (2010) The British industrial revolution: A tribute to freedom and human potential, *The Objective Standard*, pp. 47–60.

Davies T (n.d.) *Calculation of CO_2 emissions*, available at https://people.exeter.ac.uk/TWDavies/energy_conversion/Calculation%20of%20CO2%20emissions%20from%20fuels.htm, accessed 15 August 2017.

Department for Business, Energy and Industrial Strategy (DBEIS) (2017) *Energy consumption in the UK July 2017*, available at https://www.gov.uk/government/uploads/system/uploads/attachment_data/file/633503/ECUK_2017.pdf, accessed 18 August 2017.

Efficiency Vermont (2016) Electric usage chart tool, available at https://www.efficiencyvermont.com/tips-tools/tools/electric-usage-chart-tool, accessed 13 August 2017.

Energy Information Administration (EIA) (2017) *Table 7.a Electricity new generation: Total (all sectors)*, available at https://www.eia.gov/totalenergy/data/monthly/pdf/sec7_5.pdf, accessed 17 August 2017.

Erakhrumen A A (2011) Global increase in the consumption of lignocellulosic biomass as energy source: Necessity for sustained optimisation of agroforestry technologies, *International Scholarly Research Network Renewable Energy*, 2011, doi:10.5402/2011/704573.

EWEA (2012) *Factsheet*, Brussels, European Wind Energy Association, available at http://www.ewea.org/fileadmin/files/library/publications/statistics/Factsheets.pdf, accessed 29 August 2017.

Federal Ministry of Food, Forestry and Consumer Protection (2011) German forests, available at http://www.bmel.de/SharedDocs/Downloads/EN/Publications/GermanForests.pdf?__blob=publicationFile, accessed 27 August 2017.

Fehrenbacher K (2015) *China is utterly and totally dominating solar panels*, available at http://fortune.com/2015/06/18/china-is-utterly-and-totally-dominating-solar-panels/, accessed 20 August 2017.

Fialka J (2016) Why China is dominating the solar industry, *Scientific American*, available at https://www.scientificamerican.com/article/why-china-is-dominating-the-solar-industry/, accessed 20 August 2017.

Finnleo (2016) Traditional vs far-infrared sauna, available at http://www.finnleo.com/finnleo-blog/traditional-vs-far-infrared-sauna-comparison-and-contrast-by-craig-lahti, accessed 14 August 2017.

Fraunhofer Institute for Solar Energy Systems (2015) *Current and future cost of photovoltaics*, available at https://www.ise.fraunhofer.de/content/dam/ise/de/documents/publications/studies/AgoraEnergiewende_Current_and_Future_Cost_of_PV_Feb2015_web.pdf, accessed 28 August 2017.

Fraunhofer Institute (2017) Photovoltaics report, available at https://www.ise.fraunhofer.de/content/dam/ise/de/documents/publications/studies/Photovoltaics-Report.pdf, accessed 20 August 2017.

Galli A, Wiedmann T, Ercin E, Knoblauch D, Ewing B and Giljum S (2011) *Integrating ecological, carbon and water footprint: Defining the footprint family and its application in tracking human pressure on the planet*, available at http://www.oneplaneteconomynetwork.org/resources/programme-documents/WP8_Integrating_Ecological_Carbon_Water_Footprint.pdf, accessed 28 August 2017.

Gieré R and Stille P (2004) Energy, Waste and the Environment: A Geochemical Perspective, *London: Geological Society of London*.

Gimpel J (1976) *The Medieval Machine: The Industrial Revolution of the Middle Ages*, London: Victor Gollancz Ltd.

The Green Interview (2010) *The lonely mission of David Hughes*, available at http://www.thegreeninterview.com/2010/05/30/lonely-mission-david-hughes/, accessed 12 August 2017.

Grigg D B (1980) *Population Growth and Agrarian Change: An Historical Perspective*, Cambridge: Cambridge University Press.

Guardian News (2016) *Carbon emissions per person, by country*, available at https://www.theguardian.com/environment/datablog/2009/sep/02/carbon-emissions-per-person-capita, accessed 28 August 2017.

Highfield R, Jameison V, Calder N and Arnoux R (2008) ITER: A brief history of fusion, *New Scientist*, available at https://www.newscientist.com/article/dn17952-iter-a-brief-history-of-fusion/, accessed 17 August 2017.

Hughes L and Rudolph J (2011) Future world oil production: Growth, plateau, or peak? *Current Opinion in Environmental Sustainability*, 3(4), pp.225–234.

International Energy Agency (IEA) (2000) *The Future Role of Coal: Markets, Supply and the Environment*, OECD Publishing, Paris, available at http://dx.doi.org.helicon.vuw.ac.nz/10.1787/9789264180994-en, accessed 15 August 2017.

International Energy Agency (IEA) (2007) *Biofuel production*, available at https://www.iea.org/publications/freepublications/publication/essentials2.pdf, accessed 28 August 2017.

International Energy Agency (IEA) (2016) *Key world energy statistics*, available at https://www.iea.org/publications/freepublications/publication/KeyWorld2016.pdf, accessed 7 August 2017.

International Energy Agency (IEA) (2017a) Oil information: Overview, available at http://www.iea.org/publications/freepublications/publication/OilInformation2017Overview.pdf, accessed 16 August 2017.

International Energy Agency (2017b) *Electricity information: Overview*, available at https://www.iea.org/publications/freepublications/publication/ElectricityInformation2017Overview.pdf, accessed 21 August 2017.

International Hydropower Association (IHA) (2016) A brief history of hydropower, available at https://www.hydropower.org/a-brief-history-of-hydropower, accessed 18 August 2017.

Intergovernmental Panel on Climate Change (IPCC) (2014) *Climate Change 2014: Mitigation of Climate Change*, Cambridge: Cambridge University Press.

Kable (2017) The top 10 biggest geothermal power plants in the world, available at http://www.power-technology.com/features/feature-top-10-biggest-geothermal-power-plants-in-the-world/feature-top-10-biggest-geothermal-power-plants-in-the-world-2.html, accessed 20 August 2017.

Krieth F (2012) Bang for the buck: Energy return on investment is a powerful metric for weighing which energy systems are worth pursuing, *Mechanical Engineering-CIME*, 134(5), pp. 26–32.

Kumar A and Sharma M P (2017) Estimation of green house gas emissions from Koteshwar hydropower reservoir, India, *Environmental Monitoring and Assessment*, 189(5), pp. 1–12.

Lacey R and Danziger D (1999) *The Year 1000: What Life Was Like at the Turn of the First Millennium*, Boston, MA: Little, Brown and Company.

Malamina P (2013) Traditional sources, in Kander A, Malamina P and Warde P (eds) *Power to the People*, Princeton, NJ: Princeton University Press, pp. 37–80.

Malanima P (2014) Energy in history, in Agnoletti M and Neri Serneri S (eds) *The Basic Environmental History*, Cham, Switzerland: Springer, pp. 1–30.

Ministry of Business, Innovation and Employment (2015) *Energy in New Zealand 2015*, Wellington, New Zealand: Ministry of Business, Innovation and Employment.

Mir-Artigues P (2016) *The Economics and Policy of Solar Photovoltaic Generation*, de Rio, Pablo: SpringerLink, Springer-ebook.

Mosley S (2014) Environmental history of air pollution and protection, in Agnoletti M and Neri Serneri S (eds) *The Basic Environmental History*, Cham, Switzerland: Springer, pp.143–170.

Narasimham (2014) Area required for solar PV power plants, available at http://www.suncyclopedia.com/en/area-required-for-solar-pv-power-plants/, accessed 28 August 2017.

National Energy Board Canada (2016) Energy conversion tables, available at https://apps.neb-one.gc.ca/Conversion/conversion-tables.aspx?GoCTemplateCulture=en-CA, accessed 13 August 2017.

NaturalGas (2013) History, available at http://naturalgas.org/overview/history/, accessed 17 August 2017.

Nixon N (2008) Timeline: The history of wind power, *The Guardian*, 17 October 2008, available at https://www.theguardian.com/environment/2008/oct/17/wind-power-renewable-energy, accessed 20 August 2017.

O'Connor P A and Cleveland C J (2011) Energy return on investment (EROI) of oil shale, *Sustainability*, 3(11), pp.2307–2322.

Office of Energy Efficiency and Renewable Energy (n.d.a) Energy cost calculator for electric and gas water heaters, available at https://energy.gov/eere/femp/energy-cost-calculator-electric-and-gas-water-heaters-0, accessed 14 August 2017.

Office of Energy Efficiency and Renewable Energy (n.d.b) A history of geothermal energy in America, available at https://energy.gov/eere/geothermal/history-geothermal-energy-america, accessed 20 August 2017.

Olivier J G J, Janssens-Meehout G, Muntean M and Peters J A H W (2016) *Trends in global CO_2 emission: 2016 report*, The Hague: PBL Netherlands Environmental Assessment Agency.

Packer N. (2011) *A beginner's guide to energy and power*, Stafford, Faculty of Computing, Engineering and Technology, Staffordshire University, February, available at http://www.rets-projectu/UserFiles/File/pdf/respedia/A-Beginners-Guide-to-Energy-and-Power-EN.pdf, accessed 13 August 2017.

Painter D S (2012) Oil and the American century, *Journal of American History* 99(1), pp. 24–39.

Paramount (2012) The Big Bus – Trailer, available at https://www.youtube.com/watch?v=68dTwJNvE1E, accessed 17 August 2017.

Quaschning V (2015) *Statistics: Specific carbon dioxide emissions of various fuels*, available at http://www.volker-quaschning.de/datserv/CO2-spez/index_e.phparious, accessed 18 August 2017.

Rao J S (2011) *History of Rotating Machinery Dynamics*, Springer e-book, DOI 10.1007/978-94-007-1165-5.

Reefton Tourism and Cloake Creative (2017) *The town of light*, available at http://www.reefton.co.nz/reefton/155-the-town-of-light, accessed 18 August 2017.

Renewable Energy Vermont (2017) *The story of Grandpa's Knob: How Vermont made wind energy history*, available at http://www.vermontbiz.com/news/october/story-grandpa%E2%80%99s-knob-how-vermont-made-wind-energy-history, accessed 21 August 2017.

Reynolds D B (2014) World oil production trend: Comparing Hubbert multi-cycle curves, *Ecological Economics*, 98, pp. 62–72.

Rickard J (2012) ATVs on test: Versatility, fuel economy and comfort are key, *Farmers Guardian*, 6 September 2012, available at https://www.fginsight.com/vip/vip/atvs-on-test-versatility-fuel-economy-and-comfort-are-key-1192, accessed 13 August 2017.

Robinson P R (2006) Petroleum processing overview, in Hsu C S and Robinson P R (eds) *Practical Advances in Petroleum Processing*, Springer eBooks.

Rule B M, Worth Z J and Boyle C A (2009) Comparison of life cycle carbon dioxide emissions and embodied energy in four renewable electricity generating technologies in New Zealand, *Environmental Science and Technology*, 43, pp. 6406–6413.

Shepherd W and Shepherd D W (2014) *Energy Studies (third edition)*, London: Imperial College Press.

Shrinivasa U and Mukunda H S (1984) Wood gas generators for small power (~5 hp) requirements, *Sadhana*, 7(2), pp. 137–154.

Solon (2013) Erlasee, available at http://www.solon.com/global/power-plants_old/references/index.html, accessed 27 August 2017.

Sothers R B (1998) Far reach of the tenth century Eldgjá eruption, Iceland, *Climatic Change*, 39(4), pp. 715–726.

Upton J (2015) *Pulp fiction; The American trees that are electrifying Europe*, available at http://reports.climatecentral.org/pulp-fiction/2/, accessed 18 August 2017.

Vale B and Vale R (1975) *The Autonomous House*, London: Thames and Hudson.

Vale B and Vale R (2005) The all-electric house: Past and future, *International Journal of Sustainable Development*, 8(3), pp. 173–188.

Vale R and Vale B (2009) *Time to Eat the Dog?: The Real Guide to Sustainable Living*, London: Thames and Hudson.

Warde P (2014) A modern energy regime, in Kander A, Malamina P and Warde P (eds) *Power to the People*, Princeton, NJ: Princeton University Press, pp. 131–158.

White J H (1979) *A History of the American Locomotive: Its Development, 1830–1880*, New York: Dover Publications.

Wirth H (2017) *Recent facts about photovoltaics in Germany*, The Frauenhofer Institute, available at https://www.ise.fraunhofer.de/content/dam/ise/en/documents/publications/studies/recent-facts-about-photovoltaics-in-germany.pdf, accessed 22 August 2017.

Worldatlas (2017) *Countries with the highest solar water heating capacity*, available at http://www.worldatlas.com/articles/countries-with-the-highest-share-of-solar-water-heating-collectors-global-capacity.html, accessed 20 August 2017.

World Coal Association (2017) Where is coal found?, available at https://www.worldcoal.org/coal/where-coal-found, accessed 15 August 2017.

World Nuclear Association (2011) Comparison of lifecycle greenhouse gas emissions of various electricity generation sources, available at http://www.world-nuclear.org/uploadedFiles/org/WNA/Publications/Working_Group_Reports/comparison_of_lifecycle.pdf, accessed 21 August 2017.

World Nuclear Association (2016) Nuclear power in Japan, available at http://www.world-nuclear.org/information-library/country-profiles/countries-g-n/japan-nuclear-power.aspx, accessed 17 August 2017.

World Wind Energy Association (WWEA) (2017) *World wind market has reached 486GW from where 54GW has been installed since last year*, available at http://www.wwindea.org/11961-2/, accessed 20 August 2017.

World Wide Fund for Nature (WWF) (2016) *Living Planet Report 2016*, available at http://www.footprintnetwork.org/content/documents/2016_Living_Planet_Report_Lo.pdf, accessed 28 August 2017.

Wright L (1964) *Home Fires Burning*, London: Routledge and Kegan Paul.

Yao Y, Chang Y and Masanet E (2014) A hybrid life-cycle inventory for multi-crystalline silicon PV module manufacturing in China, *Environmental Research Letters*, 9(11), p. 114001 (11 pp).

YouTube (n.d.) *Human power shower – Bang goes the theory – BBC One*, available at https://www.youtube.com/watch?v=C93cL_zDVIM, accessed 14 August 2017.

Yue D, You, F and Darling S (2014) Domestic and overseas manufacturing scenarios of silicon-based photovoltaics: Life cycle energy and environmental comparative analysis, *Solar Energy*, 105, pp. 669–678.

Zabawski E (2011) History of petroleum, *Tribology and Lubrication Technology* 67(9), p. 6.

13

DWELLING

Brenda Vale, Robert Vale and Fabricio Chicca

The act of dwelling

"Dwelling" can be both a noun and a verb. We all live somewhere, and at different scales. Children understand the idea of dwelling at different scales when they write their addresses along the lines of "name, house, street, suburb, city, country, the earth, the universe". This book looks at three of these scales, the ecological footprint (EF) of the house, the EF of the country and the limits of the planet. We live on a planet where the resources on which we build our lives come from the earth. However, our identity is also formed from the country where we live; the city, town or suburb where we find the things we use on a daily basis (school, place of work, shops); and the building that we call home. This is what is covered by the idea of dwelling as a verb. The dwelling is the whole place where each of us lives and that gives us our identity, not just the building.

The German philosopher Martin Heidegger (1951) made this distinction between place and space in his essay "Building, dwelling, thinking", bemoaning the effect of modernism on the supply of dwellings, especially the effect of the modernist blocks where the shopping street and nursery school were provided in a self-contained housing block. Frampton (1974) picked up this idea, lamenting how the international use of new technologies in architecture had divorced building from place, as if we were all learning a common language like Esperanto, rather than our mother tongues.

Dwellings in the stories

The share of a dwelling in a personal EF is a very small part. In another analysis of EF using a top-down approach, the dwelling, including the materials and energy to build and operate it, formed only 4 per cent of the total EF of someone living in Wellington, New Zealand, in 2006 (Field, 2011: 108–114). In this chapter, we are only concerned with the materials of the house and the land it occupies, as energy is dealt with in Chapter 12. The dwelling's EF also depends on the number of people using it.

Comparing dwelling size

Everyone in our stories has some sort of shelter, just as everyone has something to eat. However, the size and materials of the dwellings vary widely. Table 13.1 starts by looking at the size of the houses. It is arranged by floor area, starting with the largest.

The largest house (in rural Brazil) has only the second-largest floor area per person, although the smallest shelter in UNMISS South Sudan, with its 10 occupants, does have the smallest, at 1 square metre/person. We know that houses in many parts of the developed world have been getting bigger (Khajehzadeh, 2017: 5). In New Zealand, the average floor area of new houses was 109 square metres in 1974 and had nearly doubled to 192 square metres by 2013 (Khajehzadeh, 2017: 6). In Australia, the average 100-square metre house at the start of the twentieth century had become a 245-square metre house by 2011 (McMullan and Fuller, 2015). In the United States, new houses increased in floor area from 139 square metres in 1970 to over 186 square metres in 1990, while average household size decreased from 3.1 to 2.6 persons (Miller 2012). In New Zealand, over a century, occupancy has halved from 5.2 to 2.7 (Statistics New Zealand, 2015). The modern trend, at least in some countries, is fewer people living in more space.

Comparing dwelling occupancy

Table 13.2 sets out the information by occupancy from most space per person to least, noting whether the dwelling is detached or conjoined and whether it is rural or urban. This information is important as it indicates the resources used per person for the house.

In Table 13.2, both the largest two and a group of four of the six smallest areas per person are found in rural settings. If you have money, you can use the available space in a rural area to build a large house. Some urban houses also offer plenty of space, such as the house in the gated community in Tucumán, Argentina, where we know the family are also relatively wealthy.

Obviously the number of occupants in a house will change as children grow up and leave home. The house in Oulu is a good example of this, as two adult children were living at home when the story was written, whereas normally only two people occupy the 120-square metre house, making a generous 60 square metres/person, compared with 24 square metres/person for the same-sized house in Nagoya,

TABLE 13.1 Dwellings from the stories ordered by size

Location	Dwelling area (square metres)	Tenure	No. bedrooms	No. bathrooms	House materials	Household size	Area/person (square metres)
Rural Brazil	785	Owned	9 (two for maids)	5	Reinforced concrete structure, plastered brick	5 plus 2 live-in maids	112
Urban Argentina (1)	298	?	4	4	Reinforced concrete structure, plastered brick, metal roof	3 all week 1 for 2 days/week 1 maid 5 days/week	74.5
Rural United States	288 (plus 56 flat over garage not included)	Owned	?	?	Brick, timber, vinyl cladding, concrete slab	2	144
Urban United States	210	?	3	3	Timber frame, stucco, concrete floors	4	52.5
Urban Malaysia	204	Rented	4	3	Reinforced concrete structure, plastered brick, clay tile	6	34
Urban Australia	135	?	3	1?	Flat in three-storey brick and concrete slab building	3	45
Rural New Zealand	130	Owned	3	2	Timber framed and clad	4	32.5
Village Germany	125	Owned	3	1 and sauna	Pumice block externally insulated, brick internal walls	2	62.5

Urban Finland	120	?	4	1 and sauna	Timber frame, brick clad, steel roof, concrete footings	2	60
Urban Japan	120	Owned	3	1	Reinforced concrete (structure)	5	24
Urban Brazil	101	Owned?	3	?	Apartment—concrete structure and brick infill	4	25.25
Urban Tonga	86	Owned	2	?	Timber frame, corrugated metal roof	7	12
Urban Morocco	82	Owned	1	1	Brick over the shop	4	20.5
Urban South Africa	80	Owned	2	1	Brick concrete duplex	5	16
Small town, United Kingdom	75.4	Owned	3	1	Brick cavity wall, timber frame tiled roof	3	25.1
Urban Argentina (2)	66	?	2	1	? (apartment building)	2	33
Urban Canada	60	Rented	1	1	? (condo)	2	30
Urban Mongolia	56	Owned plus ger on site for visitors	2	1	Brick, pvc, fabric and canvas roof	3	18.7
Rural Indonesia	48 and separate 12 square metres kitchen	Community owned	1	0	Lashed wood, bamboo and palm thatch	5	9.8

(*Continued*)

TABLE 13.1 (Continued)

Location	Dwelling area (square metres)	Tenure	No. bedrooms	No. bathrooms	House materials	Household size	Area/person (square metres)
Rural Myanmar	42.75		2	Latrine 10 metres from house, small structure for showers	Timber, galvanised steel roof	4	10.7
Urban Cuba	35	? Converted apartment (paid for conversion)	1 + sleeping alcove	1	?	3	11.2
Rural Mozambique	24	Materials provided by an NGO	1 (2 rooms in total)	? (not in house)	Brick, thatched roof, soil-cement floor	10	2.4
Rural India	18	Owned?	1	Pit latrine near the house, well water	Brick and timber	3	6
Urban South Sudan	16 plus separate kitchen	owned and self-built	2 (only rooms in house)	Bathing hut and communal latrine	Mud, timber, metal sheet	7	2.3
UNMISS South Sudan	10	Materials provided by an INGO	2	0	Tarps, wooden poles, nails	10	1

(1) The story from Argentina involves two dwellings; the first is the large house in a gated community.
(2) The second dwelling in the Argentina story is the grandparents' small apartment.

TABLE 13.2 Houses ranked by occupancy, location and whether they are detached or conjoined in some way

Location	Occupancy square metres/person	Detached	Conjoined	Urban	Rural
St Tammany, United States	144	√			√
Dourados, Brazil	112	√			√
Tucumán, Argentina (1)	74.5	√		√	
Eschenlohe, Germany	62.5	√			√ (village)
Oulu, Finland	60	√		√	
Celebration, United States	52.5	√ ?		√	
Sydney, Australia	45		√	√	
Penang, Malaysia	34	√		√	
Tucumán, Argentina (2)	33		√	√	
Near Stratford, New Zealand	32.5			√	√
Toronto, Canada	30		√	√	
São Paulo, Brazil	25.25		√	√	
Newton-le-Willows, United Kingdom	25.1		√	√	
Nagoya, Japan	24	√		√	
Marrakesh, Morocco	20.5		√	√	
Ulaanbaatur, Mongolia	18.7	√		√	
Johannesburg, South Africa	16		√	√	
Nuku'alofa, Tonga	12	√		√	
Havana, Cuba	11.2		√	√	
Lake Inle, Myanmar	10.7	√ ?			√
Kendeng Mountains, Indonesia	9.8	√			√
Khajuraho, India	6	√			√
Inhassoro, Mozambique	2.4	√			√
Juba, South Sudan	2.3	√		√	
UNMISS, South Sudan	1	√		√	

which provides a home for the family of two adults and three small children. In the developed world, there is a steady increase in the number of older people in the total population, and hence smaller households. In New Zealand, it is predicted that by 2050, the 65+ age group will form 25 per cent of the total population (Yavari and Vale, 2016). A similar situation has occurred in Australia (Judd et al., 2014), and in China by 2050, 25 per cent of the population is also predicted to be over 65 (Feng et al., 2012). Among the stories in this book, perhaps only rural United States, Germany and Finland fit this pattern, and the majority of the stories concern nuclear and extended families.

What is apparent in Table 13.2 is the wide difference in space per person, which can be read as the resources that go into housing each person. These resources are not necessarily used effectively. Khajehzadeh (2017: 277) found that in a 24-hour period, "regardless of house size people spend 84-100% of their time at home indoors in the core part of the house, comprising a living room, dining room, kitchen and one bedroom per occupant". Given that the majority of time spent at home for most people is in bed asleep, this suggests that much space in large houses will have very little use apart from for storing possessions. For many people in the developed world, houses are more like not-very-good museums—full of things that are not labelled and are seldom visited.

Using the space around the house

Although the space allocations are very small in South Sudan and in some rural locations, like Mozambique, Table 13.2 takes no account of space for dwelling outdoors around the house. Where climates are warm enough, dwelling, as a verb, occupies more area than the enclosed floor area of the "house"—even television can be watched outside, as in the story from Havana. The story from the Medina in Marrakesh also shows how the urban space around the house is an important part of the social life of those who live there, as is the street in the story of rural India.

Flexible space

When it comes to the use of resources, although the contribution of the house to the overall EF is small, it is obvious that some dwellings involve many more resources than others. The house in the UK is intriguing in this regard as within its modest 75.4 square metres, it has three bedrooms, a bathroom, kitchen and living space. Other small dwellings make use of flexible space, so that living rooms become spaces for sleeping at night. The house in Nagoya maintains the Japanese tradition of everyone sleeping in the same room, even with bedrooms to spare. The houses in Marrakesh and Tonga have living areas that are also used for sleeping and the very small houses and shelters are also, of necessity, used flexibly. Table 13.1 suggests that dwelling could be provided with fewer resources if the developed world could manage with smaller houses and less space to store accumulated possessions. However, the key to being able to use space flexibly has to do with possessions and storage—the fewer the possessions, the easier the question of storage.

Building your own house

Building a house single-handedly is rare compared with the community coming together to build a house, as typified in the tradition of "barn raising" in the United States (Perkins, 2004). How many people you can muster also affects what you can build, with some 200 volunteers involved in a modern barn raising in the United States (Auldridge, 2007). The small *tukul* built by Patrick for his family reflects the fact that he was the builder and also had limited money to spend on the materials. The house of the Baduy family is larger, but it is a community structure and made only of local materials. The apartment in Havana is also self-built, but there it is constrained by space as much as by materials availability and skills. Most people in the stories, however, have to buy or rent the dwelling they can afford that also offers them the most convenience. Urbanisation and development have dramatically reduced the opportunity to build for ourselves.

Building a house yourself is a way of obtaining a house for less money because of the "sweat equity" it contains in the form of your own free labour. In the past, self-building was more prevalent in developed countries. In 1927 in Stockholm, an experiment set out to make houses as cheap as the land they stood on. Timber-framed house designs were standardised, and both formwork for the concrete foundations and machines for making concrete blocks could be borrowed so that "an able bodied workman putting in his weekends, holidays and daily after-work hours in the long northern summer days, with the aid of his family and friends, could build the greater part of his own house" (Gray, 1946: 88). What emerges from the stories is that in the contemporary world, most dwellings are commodities to be marketed and purchased or rented, and that someone is making money from this method of housing people.

Commodification of the house

Heidegger (1951) was concerned with the effect of international modernism on dwelling. One aspect of the internationalisation of house construction methods is the commodification of the house, turning it from a dwelling into a consumer item. Ronald (2009) sees the modern concrete Japanese house as a commodified product although it retains aspects of traditional houses even to the extent of being perceived as impermanent, like the timber and paper houses of the past, leading to periodic demolition and rebuilding. In our story, the modern concrete house contains a traditional Japanese room on the ground floor, and the family also maintains the tradition of all sleeping on mats in one of the upstairs bedrooms, but there is also an expectation that this might change as the three girls in the family grow up and conform to the idea of having private rooms. The more the levels of privacy in the house, the more rooms that can be sold and the greater the advantage to those selling houses. This is a long way from the way people used to dwell, where sleeping together was for companionship and warmth.

House EFs

Table 13.3 sets out the EFs of each house in global hectares/person from largest to smallest, together with the ranking of the house in terms of total floor area and floor area per person. The EF, as described in Chapter 8, is the impact of the resources that go into making the house. The rankings have been adjusted to omit the grandparents' apartment in Tucumán, Argentina, which is not part of the main EF study, but which appears in Tables 13.1 and 13.2 because we know its size and features.

Most houses in Table 13.3 have reasonably consistent rankings—the very low EF houses at the bottom of the table are also small in size and offer the least floor area

TABLE 13.3 Ranking of houses by EF, overall area and area/person

Place	EF	EF ranking	Ranking house square metres	Ranking square metres/person
Rural Brazil	0.49	1	1	2
Rural United States	0.28	2	3	1
Rural Germany	0.22	3	8	4
Urban Japan	0.13	4	=9	13
Urban Malaysia	0.12	=5	5	8
Urban United Kingdom	0.12	=5	15	12
Urban United States	0.10	7	4	6
Urban Argentina★	0.08	=8	2	3
Urban Brazil	0.08	=8	11	11
Urban Tonga	0.08	=8	12	17
Urban Finland	0.06	=11	=9	5
Rural New Zealand	0.06	=11	7	9
Urban Australia	0.05	13	6	7
Urban Morocco	0.03	14	13	14
Rural Myanmar	0.02	=15	20	19
Urban South Africa	0.02	=15	14	16
Urban Canada	0.02	=15	16	10
Urban Cuba	0.01	=18	20	18
Rural India	0.01	=18	22	21
Urban Mongolia	0.01	=18	17	15
Rural Indonesia	0.00	=21	18	20
Rural Mozambique	0.00	=21	21	22
Urban South Sudan	0.00	=21	23	23
UNMISS South Sudan	0.00	=21	24	24

★House in gated community only

per person, and the two very large houses at the top of the list are also at the top for other aspects. Table 13.3 shows that apartments, whatever their size, do tend to have lower EFs than detached houses.

The act of dwelling and the stories

Heidegger (1951: 350) notes that "We do not dwell because we have built, but we build and have built because we dwell". The implication of this observation is that where we live amounts to much more than the bricks and mortar or timber and woven bamboo that we call "home". Heidegger's interest lies in the connection between the building and the place. We could extrapolate this idea and go on to say it means understanding how we live on the earth. This understanding is clear in the stories of the Baduy in Indonesia and the family living on and near Lake Inle in Myanmar. It is also clear in the story of Mozambique that the family feel the loss of dwelling caused by the uprising as they have lost their traditional way of living dependent on fishing, not because the fish are not there to catch but because there is no means of getting the fish to market. When it comes to those living in the developed world, perhaps the Finnish family come closest to the idea of dwelling, as to be Finnish is to collect berries and fruits from the forest in the autumn. Life in Finland still includes this, even if you live in a modern house. What most of the stories show us is that where people live in urban houses and apartments, it is harder to see the connection between the house and the place of dwelling. As Schumacher (1973: 109) notes, how we choose to use land and the natural resources that come from it "is not a question of what we can afford but what we choose to spend our money on". What is apparent from the stories, however, is that those who do not have the money to spend on houses come much closer to Heidegger's idea (and ideal) of dwelling than those that do.

References

Auldridge P (2007) A barn raising, *Ranch and Rural Living*, 88(4), pp. 23–26.
Feng Z, Lui C, Guan X and Mor V (2012) China's rapidly aging population creates policy challenges in shaping a viable long-term care system, *Health Affairs*, 31(12), pp. 2764–2773.
Field C (2011) The Ecological Footprint of Wellingtonians in the 1950s, MBSc thesis, Victoria University of Wellington, New Zealand.
Frampton K (1974) On reading Heidegger, in Nesbitt K (ed) (1996) *Theorizing a New Agenda for Architecture*, New York: Princeton Architectural Press.
Gray G H (1946) *Housing and Citizenship*, New York: Reinhold Publishing Co.
Heidegger M (1951) Building, dwelling, thinking, in Krell D F (ed) (1977) *Martin Heidegger: Basic Writings*, New York: HarperCollins, pp. 343–363.
Judd B, Liu E, Easthope H, Davy L and Bridge C (2014) Downsizing amongst older Australians. AHURI Final Report No.214. Melbourne, Australia: Australian Housing and Urban Research Institute.
Khajehzadeh I (2017) An investigation of the effects of large houses on occupant behaviour and resource-use in New Zealand, PhD thesis, Victoria University of Wellington, New Zealand.
McMullan M and Fuller R (2015) Spatial growth in Australian homes (1960–2010), *Australian Planner*, 52(4), pp. 314–325.
Miller B J (2012) Competing visions of the American single-family home: Defining McMansions in the *New York Times* and *Dallas Morning News*, 2002–2009, *Journal Urban History*, 38(6), pp.1094–1113.
Perkins S (2004) *Barn Bees and Others: How Collective Endeavour Built a Nation*, Swartz Creek, MI: Broadblade Press.
Ronald R (2009) Privatization, commodification and transformation in Japanese housing: Ephemeral house – eternal home, *International Journal of Consumer Studies*, 33(5), pp. 558–565.
Schumacher E F (1973) *Small Is Beautiful: Economics As If People Mattered*, New York: Harper and Row.
Statistics New Zealand (2015) *The story of the century—Dramatic changes in housing and population*, available at http://www.stats.govt.nz/Census/2013-census/profile-and-summary-reports/century-censuses-media-release.aspx, accessed 8 August 2017.
Yavari F and Vale B (2016) Alternative housing options for older New Zealanders: The case for a life-cycle study, in Zuo J, Daniel L and Soebarto V (eds) *Fifty years later: Revisiting the role of architectural science in design and practice: Proceedings of the 50th International Conference of the Architectural Science Association 2016*, Adelaide, Australia: Architectural Science Association & The University of Adelaide, pp. 527–536.

14
CONSUMER GOODS

Brenda Vale, Robert Vale and Fabricio Chicca

The rise in consumption

We are told we live in a consumer society, at least in the developed world (Flavin, 2004). This chapter looks at the history of consumer goods and how we arrived at this position.

The history of possessions

The history of consumer goods is really the history of wills, or in other words, what people leave behind. Friedman (2009: 3) makes the point well.

> When people die, everything they think they own, everything struggled, scrimped, and saved for, every jewel and bauble, every bank account, all stocks and bonds, the cars and houses, corn futures or gold bullion, all books, CD's [sic], pictures, and carpets—everything will pass on to somebody or something else.

What we learn from wills is the kinds of things people owned. Skeel (1926) pointed out that many medieval wills begin with leaving possessions such as cattle and other goods to the church as a type of insurance for a favourable reception in heaven. Apprentices were left the tools of their masters and guilds were also the receivers of valued property. Personal possessions singled out in wills were books and drinking cups and houses, furniture, cloth and clothing. These were what was needed for a civilised life at that time.

It was industrialisation that truly ushered in a consumer society. Clarke (2010: 38) argues that industrialisation meant that not only were more goods produced, but that machines meant that quality was more consistent and even better.

For the modern history of consumerism, the situation immediately after World War II was one of moving from full war production back into the domestic market. This meant ensuring there was an expanded domestic market for products, which in turn was fuelled by increased expenditure on advertising (Clayton, 2010). In 1955, the American economist Victor Lebow (1955) stated: "Our enormously productive economy demands that we make consumption our way of life, that we convert the buying and use of goods into rituals, that we seek our spiritual satisfactions, our ego satisfactions in consumption". Lawton (2013: 75) points out that the post-war surge in consumerism brought with it planned obsolescence, through cheap goods quickly wearing out or more expensive ones being replaced as fashions changed.

Private property

Possessions are private property. Willard Hurst (1964: 9) defined property as the "legitimate power to initiate decisions on the use of economic assets". Property in economic terms has the characteristics of excludability, used to distinguish between what is private and what is common, and rivalry, which refers to whether goods are exhausted when consumed (Schnegg, 2016). In a household, there is less excludability than between households, although this may also depend on cultural differences. In times of emergency, attitudes to excludability may also change with more altruistic behaviour because of being in a common adverse situation (Lemieux, 2014). However, when it comes to the environment and issues like climate change, it seems altruistic behaviour is largely missing. This leads to the issue of what is common property—something that belongs to all—and what is private in the sense that it is there to be exploited, if that is what I want to do.

Communal ownership

From the stories, it is clear that if, like the Baduy, your way of life is traditional and unchanged over many years, the community within which you live is more important than a sense of private property. Everyone in such a community understands the need to conserve the resources available to the community and not to overexploit them. As an example, in Penglipuran, which is another traditional village in Indonesia but this time on the island of Bali, the houses use local bamboo for their structure, with the thick stems halved for roof tiles, and also as woven bamboo doors. The bamboo is taken from the forest around the village but its harvesting is carefully controlled, so that this communal resource is sustained rather than exploited (Vale, 2017: 71). There is also an optimum population that the village can support in this way and, once reached, no more people can be added to the village. This is a simple fact implicit in the operation of many traditional societies but missing from the modern developed world.

In a globalised society, it is much, much harder to see the effect of taking just a bit more bamboo than last time from the forest to sell on and make some extra money. In a finite world, there has to be a limit to the resources that can be used.

This exploitation in the face of knowing that it cannot continue has been described as the tragedy of the commons. This means that in doing an action that helps ourselves—consuming goods—we fail to account for the negative effect it may have on what is common to all of us—the earth's climate. Hardin (1968) explains it well, using the example of communal grazing on common land, where the common is being grazed to its maximum sustainable capacity. Each individual herdsman sees the benefit to himself of adding another animal to his herd on the common, but the dis-benefit of overgrazing is shared between all the other herdsmen and so does not seem so important, so "the rational herdsman concludes that the only sensible course for him to pursue is to add another animal to his herd. And another; and another ... But this is the conclusion reached by each and every rational herdsman sharing a commons".

This example makes it very clear why small groups are better at husbanding resources. In a small group, you are more likely to come together to make a decision when it comes to how many animals can be grazed on the common, and the right to graze a number of animals can be passed on like property (O'Grady, 2009). Restricting grazing to those who have rights ensures common land is not overgrazed. However, when it comes to shared commons like the air, or the right to have wilderness, how is overexploitation of these commons to be regulated and policed when so very many people have a share in these commons? Hardin (1968) points out that it is cheaper for everyone to discharge their wastes into the air, water or land than to clean them up before discharge. So why should one person step out of line and bear an increased cost when no-one else does? Bringing this back to consumer goods, should the cost include not just the cost of production, transport and retail but also of ensuring that the goods are safely disassembled, recycled and reused at the end of their life? Are the resources that go into consumer goods really common property since these are the resources common to all?

Hardin's parable of the commons shows what is happening as the pressure of national governments on economic growth pushes the system to encourage people to have more and more stuff—consumer goods, holidays or cars—without thought to whether continued growth is possible within a finite system—which it patently is not. It is not possible to maximise both economic growth and environmental protection at the same time (Hardin, 1968). MacLellan (2015) goes further in his article on reinterpreting Hardin by concluding "we question the wisdom of limitless economic growth as we proceed into an uncertain and dangerous century of climate change".

It is obvious from the stories in this book that some people have a lot and some have very little. It also seems that if you are poor, you are more likely to share resources, at least within your extended family. Patrick in South Sudan built his house on land owned by his brother. The concept of ownership here is blurred, perhaps not as much as in the Indonesian examples discussed above, but it does suggest that where society is based on notions of private property it may be harder to live a low-ecological footprint (EF) lifestyle.

Estimating the impact of consumer goods

Although the size of the house in each of the stories (see Chapter 13) gives some indication of the level of income and hence the ability to acquire consumer goods, the approach taken to determining the EF of consumer goods in this book is to look only at appliance ownership as a proxy for estimating the whole EF of consumer goods in the story.

Appliances

Part of the growth in consumerism is the growth in appliance ownership. Letschert (2010) states: "most households with access to electricity will purchase a refrigerator if they can afford it". Some 75 per cent of world households own a refrigerator, but only 25 per cent in India do (Majumder, 2015). Ownership in China has risen from 24 per cent of households in 1994 to 88 per cent in 2014, "whereas in Peru, which has similar GDP per head but is more rural, it is still only 45% (Anon, 2014)". Long (2016) states that not only does virtually every home in the United States have a refrigerator, but that 23 per cent have two or more.

Cooking is common to all the stories, but the type of stove will differ. In the developed world, most households have a stove powered either by gas or electricity. In the developing world, cooking is often on a small portable stove powered by LPG (Van Leeuwen, 2017), or using wood or dried dung.

Dishwashers, however, are found only in the homes of the relatively wealthy. Around 60 per cent of households in the United States have a dishwasher, mostly higher-income households (DuPont, 2013). In a different societal context, in Indonesia those who could afford a dishwasher are likely to have paid help to wash the dishes (Euromonitor, 2017).

Globally in 2013, there were an estimated 840 million laundry-washing machines, consuming 2 per cent of the global residential sector electricity (Barthel and Götz, 2013: 4), although the authors state that ownership is uneven across the world. Most households in North America, Western Europe and the OECD Pacific countries (Australia and New Zealand) own a washing machine. Fewer people own an electric appliance for drying washing. Ownership of tumble dryers in the United Kingdom only increased from 54 per cent of households in 2001/2002 to 57 per cent in 2010 (Office of National Statistics, 2011). However, in 2009 in the United States, 78 per cent of households owned a laundry dryer, and market penetration of dryers in Europe is less than in the United States (IEA ETSAP, 2012).

When it comes to small appliances that do not use a lot of energy, the level of ownership, the type of technology and even the manufacturer become more important. Televisions are a good example, as owning multiple sets can push up the environmental impact. In 2002, Norway and Switzerland topped the world in terms of television ownership, with 99.97 per cent and 99.83 per cent of households having a set, while the fewest households with television were in the Central

African Republic and the Democratic Republic of the Congo, with 1.81 per cent and 1.69 per cent, respectively (NationMaster, 2017b). In 2009 in the United States, 82 per cent of homes had more than one TV set, and on average there were more TV sets than people in a household—2.86 TV sets and 2.5 people (The Nielsen Company, 2009). In our African stories, only the family in South Africa have a set, and then just the one, whereas the urban United States household of four people has five sets.

Table 14.1 matches 2014 average access to a working computer in selected countries (Pew Research Centre, 2017) with ownership from the stories. Rural Brazil is high, but part of this may be due to running the farming operation. Again, the averages hide wide differences in home computing. Excluding the United States, the median value for the world is 38 per cent of households with a working computer.

In 2017, the world had an estimated 4.77 billion mobile phone users, up from 4.61 billion in 2016 (Statista, 2107). Although mobile phones are not part of the analysis, they were mentioned in the stories, and of all available gadgets, the phone is the one everyone has (or wants). If nothing else, people want to be in communication.

Appliances in the stories

Table 14.2 sets out the appliance ownership for the stories. The ordering is again by dwelling size, from the largest to the smallest, to reflect a possible relationship between size of house and number of appliances. Totals are given where we are sure all or a majority of appliances have been recorded.

As might be expected, larger houses tend to have more appliances, the exceptions being Canada and the United Kingdom that manage to pack in 11 appliances into a small floor area, although these appliances are shared by two people in Canada and four in the United Kingdom. To complete the picture, Table 14.3 sets out the CO_2 emissions and EF associated with just having (but not operating) these appliances and the assumed EF for all consumer goods for each household in the stories based on the method set out in Chapter 8. The stories are ordered from largest to smallest.

TABLE 14.1 Access to a working computer at home

Country	Percentage access to computer at home	Number of computers in story
United States	80	1 (rural) 2 (urban)
Argentina	58	3
Brazil	55	7 (rural) 4 (urban)
Malaysia	51	4
South Africa	26	0
Indonesia	13	0
India	11	0

TABLE 14.2 Appliance ownership from the stories

Place	TV	Lap-top/tablet	Desktop	DVD player/ stereo	Fridge	Freezer	Stove	Microwave	Dish washer	Washing machine	Dryer	Total
Rural Brazil	2	5	2	3	1	1	1	1	1	1	1	18
Urban Argentina (1)	3	2	1		1	2	1	1	1	1	1	13
Rural United States	1*	0	1		1	1	1	1	1	1	1	9
Urban United States	5	1	1		1	1	1	1	1	1	1	13
Urban Malaysia	1	4		1	1	1	1	1	1	1		11
Urban Australia	4				1	1	1	1	1	1		10
Rural New Zealand	2	3		1	1	1	1	1	1	1		12
Urban Finland	1	2	3	2	1	2	1	1	1	1		15
Urban Japan	1		1		?	?	1	?	1	1	1**	6
Urban Brazil	3	4			1	1	1	1	1	1		13
Village Germany				1	1	1	1	1		1		6
Urban Tonga	1			1			1					4
Urban Morocco	2		1		1		1	1		1		6
Small town, United Kingdom	1	2		1	1	1	1	1	1	1	1	11
Urban Argentina (2)	2	1			1		1	1		1		7
Urban Canada	1	3			1	1	1	1	1	1	1	11
Urban Mongolia	1		1	2	1							5
Rural Indonesia	–	–	–	–	–	–	–	–	–	–	–	–
Rural Myanmar	1											1
Urban Cuba	1			1	1		1			1		4
Rural Mozambique	–	–	–	–	–	–	–	–	–	–	–	–
Urban South Africa	1				1		1	1		1		5
Rural India	–	–	–	–	–	–	–	–	–	–	–	–
Urban South Sudan	–	–	–	1***	–	–	–	–	–	–	–	–
UNMISS South Sudan	–	–	–	–	–	–	–	–	–	–	–	–

*Not mentioned in the story, but assumed, given the high TV ownership in the United States
**Mattress dryer
*** Not considered for CO_2 calculation
(1) The story from Argentina involves two dwellings; the first is the large house in a gated community.
(2) The second dwelling in the Argentina story is the grandparents' small apartment (see Table 13.1).

TABLE 14.3 Comparison of CO_2 emissions and EFs for consumer goods

Place	Household CO_2 emissions appliances tonnes/annum	Per person CO_2 emissions appliances kilograms/annum	EF appliances global hectares/person	EF consumer goods (estimated) global hectares/person
United Kingdom	0.31	100.0	0.34	2.47
Germany	0.16	80.0	0.26	1.89
Finland	0.45	225.0	0.24	1.74
Brazil rural	0.60	82.9	0.21	1.52
Japan	0.17	34.0	0.13	0.94
Argentina	0.34	44.3	0.11	0.80
United States urban	0.30	70.0	0.11	0.80
Malaysia	0.32	48.3	0.01	0.72
New Zealand	0.19	47.5	0.10	0.72
Brazil urban	0.26	65.0	0.09	0.65
Tonga	0.22	31.4	0.07	0.51
Canada	0.35	175.0	0.06	0.44
United States rural	0.19	80.0	0.06	0.44
Morocco	0.16	40.0	0.03	0.21
Australia	0.24	80.0	0.02	0.15
South Africa	0.10	2.0	0.02	0.15
Mongolia	0.10	33.3	0.01	0.07
Cuba	0.02	6.7	0.00	0.00
India	0.00	0.0	0.00	0.00
Indonesia	0.00	0.0	0.00	0.00
Myanmar	0.01	2.5	0.00	0.00
Mozambique	0.00	0.0	0.00	0.00
South Sudan (Juba)	0.00	0.0	0.00	0.00
South Sudan (UNMISS)	0.00	0.0	0.00	0.00

These results are very crude, and a more detailed study of the impact of consumer goods is needed, especially if where goods are made is taken into account, given the nature of world trade. This is where the top-down method of calculating EFs is much better since it deals with resource flows into and out of countries. Apart from asking people in the stories to check the country of origin of all their consumer goods, which seemed far too arduous, the bottom-up method fails to account for trade. In the end, the simple strategy is to have less if we want to reduce the impact of consumer goods. Remember, grave goods aside, you can't take them with you when you're gone.

References

Anon (2014) Cool developments: How chilled food is saving lives, *The Economist*, available at https://www.economist.com/news/international/21603031-how-chilled-food-changing-lives-cool-developments, accessed 24 August 2017.

Barthel C and Götz T (2013) *The overall worldwide saving potential from domestic washing machines*, Wuppertal Institute for Climate Change, available at http://www.bigee.net/media/filer_public/2013/03/28/bigee_domestic_washing_machines_worldwide_potential_20130328.pdf, accessed 24 August 2017.

Clarke M (2010) *Challenging Choices: Ideology, Consumerism and Policy*, Bristol: Policy Press.

Clayton D (2010) Advertising expenditure in 1950s Britain, *Business History*, 52(4), pp. 651–665.

designboom (2017) Giradora foot powered washer and spin dryer, available at https://www.designboom.com/design/giradora-foot-powered-washer-and-spin-dryer/, accessed 25 August 2017.

DuPont (2013) *2013 North American Dish Market Study*, available at http://fhc.biosciences.dupont.com/fileadmin/user_upload/live/fhc/DuPont_US_Dishwashing_infographic301013.pdf, accessed 26 August 2017.

Energy Information Administration (2011) *Air conditioning in nearly 100 million U.S. homes*, available at https://www.eia.gov/consumption/residential/reports/2009/air-conditioning.php, accessed 27 August 2017.

Euromonitor (2017) Dishwashers in Indonesia, available at http://www.euromonitor.com/dishwashers-in-indonesia/report, accessed 26 August 2017.

Flavin C (2004) Preface, in L Stark (ed.) *State of the World 2004: A Worldwatch Institute Report on Progress Toward a Sustainable Society*, New York: W.W. Norton and Company, Inc., pp. xvii–xix.

Friedman L M (2009) *Dead Hands: A Social History of Wills, Trusts, and in Heritance Law*, Stanford, CA: Stanford Law Books.

Hardin G (1968) The tragedy of the commons, *Science, New Series*, 162(3859), pp. 1243–1248.

International Energy Agency Energy Technology Systems Analysis Programme (IEA ETSAP) (2012) *Dryers*, available at https://iea-etsap.org/E-TechDS/PDF/R09_Dryers_FINAL_GSOK.pdf, accessed 26 August 2017.

Lawton M (2013) Consumer goods, in Vale R and Vale B (eds) *Living within a Fair Share Ecological Footprint*, London: Earthscan, pp. 73–83.

Lebow V (1955) Price competition in 1955, *Journal of Retailing*, available at http://www.gcafh.org/edlab/Lebow.pdf, accessed 4 January 2018.

Lemieux F (2014) The impact of a natural disaster on altruistic behaviour and crime, *Disasters* 38(3), pp. 483–499.

Letschert V (2010) Material world: Forecasting household appliance ownership in a growing global economy, Lawrence Berkeley National Laboratory, LBNL Paper-2403E, available at http://escholarship.org/uc/item/28h0w89d, accessed 24 August 2017.

Long H (2016) 23% of American homes have 2 (or more) fridges, available at http://money.cnn.com/2016/05/27/news/economy/23-percent-of-american-homes-have-2-fridges/index.html, accessed 25 August 2017.

MacLellan M (2015) The tragedy of limitless growth: Re-interpreting the tragedy of the commons for a century of climate change, *Environmental Humanities* 7(1), pp.41–58.

Majumder S (2015) *The village that just got its first fridge*, available at http://www.bbc.com/news/magazine-30925252, accessed 25 August 2017.

NationMaster (2017) Media-households with television: Countries compared, available at http://www.nationmaster.com/country-info/stats/Media/Households-with-television, accessed 25 August 2017.

The Nielsen Company (2009) *More than half the homes in U.S. have three or more TVs*, available at http://www.nielsen.com/us/en/insights/news/2009/more-than-half-the-homes-in-us-have-three-or-more-tvs.html, accessed 26 August 2017.

Office of National Statistics (2011) Ownership of consumer durables increases into 2010, available at http://webarchive.nationalarchives.gov.uk/20160129163323/http://www.ons.gov.uk/ons/dcp171780_245268.pdf, accessed 26 August 2017.

O'Grady E (2009) Hall v Moore and others, *The Estates Gazette*, p. 99.

Pew Research Centre (2017) *Global computer ownership*, available at http://www.pewglobal.org/2015/03/19/internet-seen-as-positive-influence-on-education-but-negative-influence-on-morality-in-emerging-and-developing-nations/technology-report-15/, accessed 26 August 2017.

Schnegg M (2016) Collective foods: Situating food on the continuum of private-common property regimes, *Current Anthropology*, 57(5), pp. 683–689.

Skeel C A J(1926) Medieval wills, *History*, 10(40), pp. 300–310.

Statista (2017) *Number of mobile phone users worldwide from 2013 to 2019 (in billions)*, available at https://www.statista.com/statistics/274774/forecast-of-mobile-phone-users-worldwide/, accessed 26 August 2017.

Vale B (2017) Building Materials, in Petrovic E K, Vale B and Pedersen-Zari M (eds) *Materials for a Healthy, Ecological and Sustainable Built Environment: Principles for Evaluation*, Cambridge: Woodhead Publishing.

Van Leeuwen R (2017) *Increasing the use of liquefied petroleum gas in cooking in developing countries*, World Bank, available at https://openknowledge.worldbank.org/handle/10986/26569, accessed 26 August 2017.

Willard Hurst J (1964) *Law and Economic Growth: The Legal History of the Limber Industry in Wisconsin, 1836–1915*, Cambridge: Harvard University Press.

15
CONCLUSIONS

Brenda Vale, Robert Vale and Fabricio Chicca

Why we wrote this book

This book looks at the daily routines behind the ways that people live and attempts to make a comparison of their impact. It has drawn on and adapted methods suggested by others for how to calculate these impacts. The results are incomplete, as there is no way of including each person's share of the impact of living in their particular country. The resulting numbers are a snapshot of life, and like all ecological footprint (EF) investigations, they are a snapshot of the past.

Doing the same thing

The stories in this book are set in very different locations and climates, and yet their fundamental similarity is startling. Life for everyone consists in getting up, organising breakfast, getting children to school (or work) and adults to work and then collecting everyone together back at home at the end of the day, for eating and sleeping. There are times in the week set aside for leisure activities, generally at the weekend. The impacts of this apparent sameness are very different between the different stories, but obviously impact has a very strong relationship to available income. This is the tragedy of the EF—the very lifestyles which would be sustainable for all of humanity are those, with exceptions, that people are anxious to escape. The story exceptions are the Baduy of Indonesia, who have chosen to eschew modern life and its accoutrements; the family living by Lake Inle in Myanmar, although they have an outboard motor for their boat rather than an oar; and the Indian family, who have very little and live with dignity. Another significant fact is that these three lifestyles are rural, and as discussed in Chapter 1, the world is urbanising.

There are two urban stories in which people are living very low-EF lifestyles—those of Cuba and Morocco. Cuba's average EF sits within the fair share, meaning that everyone could have the lifestyle of those living in Cuba (Rees and Moore, 2013: 17). For those who promote the compact city (Jenks et al., 1996; Newman and Kenworthy, 1989; Jenks and Burgess, 2000) the story from Morocco shows how compaction reduces EF as the family can walk everywhere because economic life is found within the Medina and there is no need to travel. But tourists come and it is their presence that at least partially provides our family with their income. Low-EF living means changing the economic basis of each place so that local economies support local communities, as Schumacher (1973) discussed in *Small Is Beautiful: A Study of Economics as If People Mattered*. He recognised that the way to change society was to reorganise its economic systems. However, the story of life in the Medina offers a view of what it is like living with everything within walking distance.

The disruption caused by war and civil unrest is also clear in the stories from South Sudan and Mozambique. This is not just physical disruption, which is bad enough, but also the way livelihoods are destroyed, and consequently a whole way of life.

Life inevitably comes back to economics. What the stories show is that modern free-market economics tend to produce ways of living that the eco-systems of the earth can no longer support. These ways of life are possible because of the exploitation of stored fossil-fuel resources to support them. We know they should not be used because of climate change, but current economic systems have locked us into using them. If nothing else, the stories emphasise the need to rethink economics if we want to live lifestyles that can be sustained into the future.

Well-being

High EFs come from high consumption. One approach to working out what people really need has been the series of well-being indicators, such as the Gross National Happiness index (Fleurbaey and Blanchet, 2013: 160). When it comes to sensory inputs, science has shown that people work on a logarithmic scale. Two trumpets sound twice as loud as one trumpet, but to double the perceived sound level again requires four trumpets, and doubling it again would require eight. Similar principles may apply when it comes to psychological satisfaction.

A similar effect of more not necessarily leading to greater happiness has been observed when it comes to gross domestic product (GDP). Easterlin observed the lack of a relationship between national GDP and reported happiness (Easterlin, 1974). In a later study (Easterlin, 2009), he found that although satisfaction rose with GDP, it was satisfaction with material things, and it came at the expense of satisfaction with the work/life balance and with health. Field (2013) reached a similar conclusion in her comparative analysis of the EF of people living in Wellington and surveys of how people felt about living in New Zealand in 1956 and approximately 50 years later (Table 15.1).

In Wellington, the 43 per cent rise in EF only produced a small rise in positive quality of life and a greater reduction in satisfaction with the balance between work and the rest of life.

TABLE 15.1 EFs and well-being in Wellington, New Zealand

	1956	2008	Difference
Positive quality of life	86%	95%	+9%
Satisfaction with work/life balance	96%	73%	−13%
Ecological footprint	4.2 global hectares	6.0 global hectares★	+43%

★Calculated for 2006

What makes for happiness is much more complicated than just comparing material things. Graham (2011) reports how "very poor and destitute respondents report high or relatively high levels of well-being, while much wealthier ones with more mobility and opportunities report much lower levels of well-being and greater frustration with their economic and other situation". Nevertheless, following the 1972 announcement that Bhutan was going to measure Gross National Happiness, The European Commission eventually took up the idea in 2009 and in 2010, The Canadian Index of Well-being was established (Zuzanek, 2013). Rather like EF and CF, that are affected by where you live, happiness is also at least in part culturally determined (Graham, 2008).

Sufficiency living

The late and highly respected King of Thailand Rama IX applied Buddhist ideas that life should be ruled by moderation and knowing what is enough to both the economy of the country and to its agriculture in the policy of Sufficiency Economy (Hengrasmee, 2013: 201–202), which led to the New Theory Agriculture (Noy, 2011), which meant a rural household could feed itself and be protected from natural phenomena such as droughts, but could still raise some crops for sale. These ideas were to be integrated into the whole economy to ensure that all Thai people could "live within what one could have and not be extravagant" (Hengrasmee, 2013: 202).

This is not to suggest that to live a low-EF lifestyle we have to be Buddhists, but rather that for low EFs, we have to grapple with the idea that we cannot have everything. It is rather the reverse; we can have very little that uses material resources. We can have as much as possible of non-material things, such as love and friendship, which could be a good thing (James-Edgar, 2015).

Happiness and EF

Table 15.2 compares the EFs for each of the stories with the happiness index, where known, of the countries in which the stories are set. The six things that make up the measure of happiness are income, life expectancy in reasonable health, being able to count on someone when things get rough, freedom and absence of corruption both in commerce and politics (Helliwell et al., 2017). Table 15.2 is ordered from highest to lowest rank in the happiness index. The final column gives the 2016

TABLE 15.2 Comparison of EF, Happiness Index and GDP

Place	EF global hectares	Happiness Index	GDP per capita US$
Finland	3.81	7.47	43,429
Canada	4.03	7.32	42,319
New Zealand	8.84	7.31	–
Australia	3.84	7.28	51,593
United States rural	4.30	6.99	57,294
United States urban	4.11	6.99	57,294
Germany	10.47	6.95	–
United Kingdom	7.63	6.71	40,412
Brazil rural	5.90	6.64	8,587
Brazil urban	2.31	6.64	8,587
Argentina	3.43	6.60	12,425
Malaysia	4.38	6.08	9,546
Japan	2.06	5.92	37,304
Indonesia	0.03	5.26	3,636
Morocco	2.67	5.24	–
Mongolia	4.94	4.96	–
South Africa	1.64	4.83	5,018
Mozambique	0.13	4.55	–
Myanmar	1.10	4.55	–
India	0.34	4.32	–
South Sudan Juba	0.10	3.59	–
South Sudan UNMISS	0.17	3.59	–
Tonga		N/A	–
Cuba		N/A	

GDP per capita in US$ for countries that fall within the top 50 world economies as ranked by GDP (StatisticsTimes, 2016).

The results for Brazil suggest that you do not have to have a high GDP per person to feel quite happy, and from the story in São Paulo (Brazil urban), you do not need to have a high EF to feel quite happy. The low GDP for Brazil is partly due to rising poverty (Phillips, 2017). In spite of this, the happiness index for Brazil is only just below that of the much wealthier but apparently morose population of the United Kingdom. The comparison between Japan and Indonesia is also worthy of note, as they have happiness indices of 5.92 and 5.24, respectively, but the per capita GDP in Japan is over 10 times that of Indonesia.

God is in the details

By now, it is clear that life as we know it in the developed world is not going to form a model for fair earth share living. The simple answer is that we have to learn to live with fewer resources. If you are poor, this is what you do, and this is what the poor have always done. Whether the future will be rural or urban is a more difficult question to answer. What has to be stopped is allowing urbanisation to spread and consume even more of the remaining productive land.

What is apparent from the stories is it is not being urban or rural that leads to a low EF, but the resources you use. If you are very poor, then a rural life, for all its hardships, may offer a better chance of supplying basic needs. This was what the New Theory Agriculture in Thailand aimed to do. If you have very few resources in an urban situation, things may be more difficult but living goes on. The biggest asset of the poor is the family, and therein lies the paradox of humanity. We know that the biggest threat to the size of the individual fair share of resources is the ever-expanding human population, but unless an equitable means of sharing resources is evolved, so that those in the developed world have a lot less, the only possible response of the very poor will be to increase numbers in order to have as many hands as possible to assist in working their way out of poverty.

Accepting the need for a dramatic adjustment to high-EF lifestyles so they become low-EF ones is one thing, but doing it will be a process that is neither easy nor comfortable. The issue, as discussed above, is the need to change economic systems and this cannot be achieved through individual or even community actions. Ultimately, many people in the developed world do nothing because the problem is too remote from them. Voting in an election once every 3 or 5 years is not the way to start changing the economic system, but in modern democracies this is the only say most people have. Moreover, the problem of overexploitation of resources is psychologically distant. It is hard to see the effects in everyday life of overexploitation of eco-system services. As Milfont et al. (2014) found in their study of New Zealanders, those living near the coast were more likely to believe in the effects of climate change and support greenhouse gas reduction because they are psychologically closer to the problem. This might suggest that tourism could be a way of bringing people closer to environmental issues. However, tourists, despite being offered tours of slums where they go explicitly to see poverty (Frenzel, 2014), do not seem to transfer the lesson of how to relieve poverty to their own lives. The connection between poverty and plenty (or even having too much) is hard to see.

Last words

We did not set out to write a depressing book. In fact, the stories are full of hope since many of them illustrate the resilience of people in coping with less-than-perfect conditions in both the developing and developed worlds. However, what has emerged from this investigation of environmental impact, as measured by the EF, is that there is an easy answer to living within a fair-share EF: consume fewer resources. There is no way that those living in poverty in parts of the world can have more unless those that already have much decide it is time that they had a lot less. This will mean a dramatic change to the way developed world societies are organised. It will mean steady-state economics and local economics ("as if people mattered", to cite Schumacher [1973]), and moving away from the idea that economic growth is the only goal a society can or should have.

References

Easterlin R A (1974) Does economic growth improve the human lot? Some empirical evidence, in David P A and Reder M W (eds) *Nations and Households in Economic Growth: Essays in Honor of Moses Abramovitz*, New York: Academic Press, pp. 89–125.

Easterlin R A (2009) Lost in transition: Life satisfaction on the road to capitalism, *Journal of Economic Behavior and Organization*, 71(2), pp. 130–145.

Field C (2013) A study of Wellington in the 1950s, in Vale R and Vale B (eds) *Living Within a Fair Share Ecological Footprint*, London: Earthscan, pp. 159–181.

Fleurbaey M and Blanchet D (2013) *Beyond GDP Measuring Welfare and Assessing Sustainability*, New York: Oxford University Press.

Frenzel F (2014) Slum tourism and urban regeneration: Touring inner Johannesburg, *Urban Forum*, 25(4), pp. 431–447.

Graham C (2008) *Happiness Around the World: The Paradox of Happy Peasants and Miserable Millionaires*, Oxford: Oxford University Press.

Graham C (2011) Does more money make you happier? Why so much debate? *Applied Research in Quality of Life*, 6(3), pp. 219–239.

Helliwell J, Layard R and Sachs J (2017) Online data, in *World Happiness Report 2017*, New York: Sustainable Development Solutions Network, available at http://worldhappiness.report/ed/2017/, accessed 29 August 2017.

Hengrasmee S (2013) A study of suburban Thailand, in Vale R and Vale B (eds) *Living Within a Fair Share Ecological Footprint*, Abingdon: Earthscan, pp. 201–214.

James-Edgar K (2015) A life full of friends, *Vibrant Life*, 31(1), pp. 12–16.

Jenks M, Burton E and Williams K (eds) (1996) *The Compact City: A Sustainable Urban Form?* London: E and FN Spon.

Jenks M and Burgess R (eds) (2000) *Compact Cities: Sustainable Urban Forms For Developing Countries*, London: Spon.

Milfont T L, Evans L, Sibley C G, Ries J and Cunningham A (2014) Proximity to coast is linked to climate change belief, *PLoS ONE*, 9(7), e103180.

Newman P and Kenworthy J (1989) *Cities and Automobile Dependence: An International Source Book*, Aldershot: Gower.

Noy D (2011) Thailand's sufficiency economy: Origins and comparisons with other systems of religious economics, *Social Compass*, 58(4), pp. 593–610.

Phillips D (2017) 'People are getting poorer': Hunger and homelessness as Brazil crisis deepens, *The Guardian*, 19 July 2017, available at https://www.theguardian.com/global-development/2017/jul/19/people-getting-poorer-hunger-homelessness-brazil-crisis, accessed 30 August 2017.

Rees W E and Moore J (2013) Ecological footprints and urbanisation, in Vale R and Vale B (eds) *Living Within a Fair Share Ecological Footprint*, London: Earthscan, pp. 3–32.

Schumacher E F (1973) *Small Is Beautiful: Economics As If People Mattered*, New York: Harper and Row.

StatisticsTimes (2016) Projected GDP Ranking (2016–2020), available at http://statisticstimes.com/economy/projected-world-gdp-ranking.php, accessed 29 August 2017.

World Bank (2017) Brazil, available at https://data.worldbank.org/country/brazil, accessed 30 August 2017.

Zuzanek (2013) Does being well-off make us happier? Problems of measurement, *Journal of Happiness Studies*, 14(3), pp. 795–815.

INDEX

Africa 14–34, 76, 111, 139
agriculture 1, 2, 3, 4, 21, 38, 40, 94–5, 160, 162, 229; urban agriculture 77
air conditioner(s) 17, 25, 52–3, 80–1, 87–8, 96, 116, 118, 119, 122
Ancient Greece 4, 7, 188
Andes 111,112
animal skins 54
apartment 67, 68, 81,84, 87, 96, 115, 117, 122, 124, 209, 211–12, 215–17, 222
appliances 25, 29, 49, 52, 68, 70, 79, 80, 87, 96, 102, 106,119, 120, 122, 134–6, 148–50, 221, 227, 244
Athens 3
Argentina 111–18, 135, 142–3, 157, 175, 180, 183, 198, 201, 209–11, 213, 215–16, 222–4, 230
Asia 2, 14, 35–62, 76, 111, 139, 188; Asian Tsunami 38
Australia 2, 91–101, 135, 142–5, 148–50, 156–8, 161–3, 165, 182–4, 198, 209–10, 213, 216, 221, 223–4, 230
aviation 107, 179

Baduy 37, 41–5, 199, 214, 216, 219, 227
Bahrain 139–40
barbecue 97, 101, 104–5, 108, 115–16, 121–2
barter(ing) 21–2, 43, 85–6
Beijing 3
berry picking 64, 72
bicycle(s) 27–8, 42, 58, 66–7, 86, 90, 117, 180–1, 189, 202; shop 68

bio-capacity 7–9, 14–15, 36, 93–4, 112, 127, 140, 143–4, 146–7, 162–3
biodiesel 174
biodiversity 9, 14
biofuels 133, 173–4, 189, 192, 197
biomass 131, 133, 189, 194–5, 198
bodega 83
boat 20–1, 58–9, 86, 180–1, 183, 227; ferry boat 170–1
Bondi *see* Sydney
bottom-up 10,224
Brazil 2, 111–14, 118–24, 129, 135, 142–4, 146, 148–50, 154, 157–8, 161–3, 165, 176, 180, 183–4, 190, 195, 198, 201–2, 209–11, 213, 216, 222–4, 230
built-up land 10, 127, 139
bus 47, 52, 55, 66, 82, 103, 106, 117, 123, 169–71, 175–6, 180–2

Canada 2, 76–81, 135, 142–6, 148–50, 157–8, 161–3, 165, 181, 183, 189, 195, 198, 201, 211, 213, 216, 222–4, 230
canvas 54
carbon dioxide (CO2) emissions 69, 73–4, 126–8, 130, 132–6, 142–3, 148–50, 155, 160–3, 167, 169–74, 179–80, 183, 191, 193, 196–7, 201, 224
carbon dioxide (CO2) sequestration 127, 130, 135, 174
carbon footprint 10, 86, 94, 160, 199
carbon neutral 14, 173, 195
Cardiff 3, 7, 162
cash crops 43, 58

cassava 107–8; *see also* manioc
Celebration 78, 87–91
central heating 72, 80, 87
charcoal 19, 21–2, 28, 31, 33, 43, 188
Choctaw 78
church 20, 24–5, 27–8, 89, 91, 108, 175, 218
civil war 15, 29, 147
climate change 2, 8, 10, 15, 147, 171–2, 201, 219–20, 228, 231
coal 8, 39, 49, 54, 67, 87, 131–3, 142, 157, 168–9, 189–94, 197–8, 201; coal equivalent 132; coal gas 193; coal heater 54–5
cohousing 148
Commodity Balance 128–9
compost 43, 48, 71
condominium 78–80
copper 5–6
consumption patterns 9, 48
consumer goods 7–8, 11, 48, 52, 67, 116, 134, 136, 142, 147, 149–50, 199, 201, 218–26
consumerism 11, 219, 221
Corinth 4
couscous day 19
Cragside 195
cropland 127–9, 140, 161, 173–4
Cuba 2, 7, 71, 81–4, 91, 135, 142–4, 146, 148–50, 154, 157–8, 161–3, 181, 183, 198, 201, 212–13, 216, 223–4, 227–8, 230
cycling 28, 50, 58, 64, 168–9, 170, 189

dairy products 56, 65, 66, 72, 146, 159, 160, 163, 165, 199
day care centre 51, 52
desert 122, 145
diabetes 51
diesel 66, 68, 80, 96, 101, 121, 170, 173, 175–7, 179–82, 193
Disney 78
district heating 67–8
dog(s) 68, 69, 85, 114, 115
domestic energy 7, 8, 134, 144, 197, 202
Dourados 2, 113, 118–22, 176, 213
drought 15, 20–1, 229
DVD 27, 29, 52, 54, 106, 189–90, 223

earth overshoot day 8
ecosystems 160
education 20, 65, 141, 147
electric cars 168–9, 172
electricity 17, 20, 25, 29, 40, 42, 45, 49, 50, 52–4, 57, 67–9, 71, 80–1, 96, 102, 116, 119–20, 131–4, 143, 158, 168–70, 181–2, 191–9, 201, 221
en-suite 71, 102
environmental impact 2–4, 7, 11, 39, 126–7, 129–31, 134, 136, 138, 141–8, 152, 154–5, 160–5, 167, 184, 195, 197, 221, 231
equity 165, 215
Eschenlohe 2, 64, 68–70, 213
ethanol 122, 174, 176–7, 180, 194, 199, 200

fair (earth) share 7, 14, 36, 76, 112, 146, 230
fan 17, 49, 81, 106, 116, 119
fast food 47, 51, 80, 88–91, 104, 129–30, 145, 162, 163, 165
favelas 114
feedlot 5
fertiliser 3, 43, 158
Finland 2, 63–8, 132, 133, 135, 142–5, 148–50, 157–8, 161–3, 181, 183, 194, 198–9, 201, 211, 213, 216–17, 223–4, 230
firewood (wood for burning) 33, 42, 44, 67–8, 132, 144,188–9, 195
fishing 15, 21–2, 38, 44, 58, 64–5, 86, 94, 216; grounds 127
flat *see* apartment
floating gardens 39, 58–9
flood(ing) 64, 84–6, 195; flood plain 114
flying 66; *see also* aviation
food chain 152
food miles 153–5
football 19, 90
fossil fuels 8, 127, 141, 147, 149, 159, 189, 192–4, 196–7, 199
free market 9,173, 197, 228
freight transport 172–3
Fukushima 49, 194
fusuma 49
futon 49
fuel consumption 59, 167, 169–70, 172, 174–9

garden 3, 20–1, 45–6, 48, 65, 71–2, 77, 85–6, 119; *see also* floating gardens
gas stove 57; hob 71; *see also* natural gas
gated communities 114, 118
GDP 94, 221, 228, 230
generator 29, 33, 87, 121; village generator 57
geothermal heating 68; power 133, 189, 194–7
ger 39, 54, 56–7
Germany 2, 63–4, 68–70, 132, 135, 140–4, 148–50, 157–8, 161–3, 181, 183, 190, 195–6, 198–202, 210, 213, 216, 223–4, 230

GHG (greenhouse gas) emissions 10, 134, 152, 158–60, 167, 191–2, 195, 197, 199
global hectare definition 9, 126–8, 138, 173
grazing land 127–9
Gwynedd 3
gym 24, 66, 79, 86, 118, 120, 123

Havana 2, 77, 81–4, 202, 213–15
HDI 140–1, 145–7, 149
heat pump 49, 102
homework 17, 19, 23, 46, 56, 69, 82, 90, 103
Hong Kong 35
hunting 64, 65, 85
hurricane 77–8; Katrina 84–7, 154
hybrid car(s) 90, 168, 170
hydroelectric power (hydroelectricity) 39, 49, 80, 116, 131, 133, 168, 195
hypermarket 51–3

India 1, 2, 36, 37, 39–41, 135, 141–4, 146, 148–50, 154, 157–8, 161–4, 180, 183, 191, 196, 198, 201, 212–14, 216, 221–4, 230
Indonesia 2, 35–6, 41–5, 135, 141–4, 146, 148–50, 154, 157–8, 161–4, 180, 183, 198–9, 201, 211, 213, 216, 219–24, 227, 230
industrial revolution 9, 11
infrastructure 8, 39, 68, 127, 131, 138–40, 143, 146, 149, 201
Inhassoro 2, 15–16, 19–22, 213
irrigation 155

Japan 2, 36–8, 45–50, 135, 142–4, 146, 148–50, 157–8, 161–3, 180, 183–4, 190, 193, 196, 198, 201–2, 211, 213–16, 223–4, 230
Johannesburg 2, 15–16, 22–5, 174–5, 213
Juba 2, 16, 26–33, 135, 143–4, 148–50, 201, 213, 224, 230

Kendeng Mountains 2, 37, 41, 213
kerosene 49, 192, 198
Khajuharo 2, 37, 39–41, 213,
kindergarten 46–8
kobhz 17

Lake Inle 2, 39, 57–60, 213, 216, 227
Lake Ontario 77
Lake Pontchartrain 78
latrine 27, 30, 40, 57, 212
leftovers 24, 48, 60, 72, 80, 82–3, 88, 105, 107, 115, 120–2
liquefied natural gas (LNG) 49

livestock unit 129
living standard 9, 202
load shedding 25
local food 153–4
London 3, 95, 155, 190

Macao 35
Malaysia 2, 36, 38, 50–3, 135, 142–5, 148–50, 157–8, 161–3, 181, 183, 198, 201–2, 210, 213, 216, 222–4, 230
manioc 21–2, 121–2; *see also* cassava
market 4, 9, 21, 28, 32–3, 83, 86, 107–8, 123, 216, black market 83, dairy market 102, domestic market 219; evaluation 119, 121; gardens 3; penetration 221; share 45; signals 174; town 64, 70; value 53; *see also* domestic market; free market; hypermarket; supermarket
Marrakesh 2, 15–19, 213–14
mate 115
mattress dryer 49
meat 19, 24–5, 28, 31–2, 44, 56, 66, 70, 72, 80, 83, 86, 88–9, 95, 97, 101, 104–5, 107, 115–16, 120–3, 128–30, 145–6, 157–65, 190
Medina 15–19, 214, 228
meditation 69
milk 5, 18, 31–2, 40, 47, 50–1, 56, 72, 83, 88–9, 98–100, 102–5, 115, 119–20, 128–9, 145, 157–60, 162–5; coconut milk 60, 107–8, 121, 123; vegan milk substitute 66
Mongolia 2, 36, 38–9, 53–7, 135, 141–5, 148–50, 154, 157–8, 161–3, 165, 181, 183–4, 198, 201, 211, 213, 216, 223–4, 230
monsoon 37
Morocco 2, 15–19, 135, 142–5, 148–50, 157–8, 161–3, 165, 180, 183, 198, 201, 211, 213, 216, 223–4, 227–8, 230
motorbike *see* motorcycle
motorcycle 28, 33, 47, 51–2, 66, 180–2
Mozambique 2, 14–16, 19–22, 135, 141–4, 146–50, 154, 157–8, 161–3, 180, 183, 198, 201, 212–14, 216, 224, 228, 230
Myanmar 2, 36, 39, 57–60, 135, 141–4, 146, 148–50, 154, 157–8, 161–3, 181, 183, 198, 201, 212–13, 216, 223–4, 227, 230

Naadam festival 56
Nagoya 2, 37–8, 45, 213–14
natural gas 50, 71, 116, 131–3, 168, 189, 192–3, 197, 201; *see also* liquefied natural gas

neighbours 17–18, 22, 55–6, 60, 82, 85–6, 105, 147
New Theory Agriculture 229, 231
Newton-le-Willows 2, 64, 70–4, 213
New Urbanism 78
New Zealand 2, 51, 93–5, 101–6, 126, 135, 138, 141–5, 148–50, 157–8, 161–3, 165, 171–2, 182–3, 190, 195, 197–8, 201, 209–1, 213, 216, 221, 223–4, 228–31
Nomadic country 38; culture/heritage 55–6, 162
non-renewable electricity 131, 199; energy 2; *see also* fossil fuels
nuclear electricity 116, 133, 197; energy (power) 50, 69, 87, 189–90, 193–4
Nuku'alofa 2, 95–6, 105–8, 213
nursery (school) 71, 73, 208

off-peak 80
oil 2, 8, 49, 131–3, 168, 172–4, 183, 189–93, 197–8, coconut oil 42, olive oil 17, 65, 83, 88; soya oil 83, 89; heaters/heating 49–50, 54, 67–9, 102, 143; reserves/resources 16, 29, 191; supply 167
Oulu 2, 63–8, 199, 209, 213
outboard motor 183–4, 227

paddy field 38, 43
Paung Daw Oo festival 58–9
Penang 2, 38, 50–3, 213
people mover 51–2
petrol 25, 47, 72–4, 96, 101, 167–9, 174–7, 179–82, 193–4
petroleum products 130, 180, 192
pets 22, 28
photovoltaics (PV) 131, 133, 196–7, 199, 200
population 1, 3–9, 14–16, 31, 35–9, 41, 45, 50, 63–4, 69, 76–8, 93–6, 111–14, 127, 139–41, 146–8, 161, 164–5, 171, 173, 188, 190–1, 213, 219, 230–1
potato 4, 6, 25, 28, 40, 58, 66, 69, 73, 80, 86, 88–9, 98–100, 104–5, 115, 120, 155, 157–60; sweet potato 21–2, 25, 43–4, 47–8, 83, 107–8
poverty line 14, 112
public transport 27–8, 32, 47, 52, 66, 79, 90, 168–72

quad bike 86, 102–3, 182, 190

radioactivity 47
rail 155, 170–3
railroad 194

railway 47, 64, 67–8, 73, 97, 140, 199
rain(fall) 15, 21, 38–9, 43, 64, 72, 77–8, 95–6, 103, 113–14, 144, 162
rainforest 112, 146
rain water 40, 106, 119
ration book 82–3
recyclable waste 3, 71
recycled goods 220; paper 68; wood 84
recycling 50; bins 68
Reefton 195
refugees 26, 68
renewable energy 143, 168, 194–7, 199
rice 22, 25, 28, 32, 37, 40, 43–4, 47–8, 50–2, 56, 58–60, 72, 83, 88–9, 99, 105, 120–3, 158, 163
Rome (Roman Empire) 4–5

St. Tammany Parish 2, 78, 84–7
São Paulo 1–2, 113–14, 122–4, 176, 201, 213, 230
sauna 65–7, 69, 190, 199, 210–11
sawmill 87
school(ing) 4, 17–20, 22–9, 31–2, 40–1, 46–7, 51–6, 58, 66–8, 82, 88, 90, 103–4, 119–20, 123, 149, 175–6, 178, 208, 227
self-driving vehicle 172
self-sufficiency 86
septic tank 87, 102, 119
sewage 3, 87, 102, 119
ship(ping) 3, 130, 154–5, 167, 172–3, 180, 193, 195
shopping 19, 25, 46, 58, 68–9, 73, 103, 117, 124, 175; bags 48; centres 79; street 208
Singapore 35, 139–40
Smith, Adam 9
snack(s) 17, 24–5, 47–8, 51, 59–60, 66, 72, 80, 86, 88, 90, 97–100, 115, 121, 123
solar: battery pack 29; collector 70; energy 8, 116, 133, 188–9, 194–8; generator 87; lights 87; panel 29, 120; power 33; water heater 106, 142, 196, 199; *see also* photovoltaics
sorghum 32
South Africa 2, 14–16, 22–5, 135, 141–4, 146, 148–50, 154, 157–8, 161–3, 174–5, 180, 183–4, 198, 201–2, 211, 213, 216, 222–4, 230
South Sudan 2, 15–16, 26–34, 135, 142–4, 146–50, 154, 161–3, 180, 183, 190, 198, 201, 209, 212–14, 216, 220, 223–4, 228, 230
soap 28, 43, 83
soy (soya bean) 44, 65–6, 113, 118–21, 157–9, 178–179

sports 11, 67, 103; fields 77
Stanley 14
Stratford 2, 95, 101–5, 213
street-car 79
storm water 3
subsistence 40, 94, 141
sufficiency economy 229
sugar cane 112, 118, 122, 176, 194; *see also* ethanol
summer cottage 65
Sundanese 41
supermarket 47–8, 72, 86, 89, 91, 97, 101, 103, 117, 121, 124
surfing 96–7, 101
SUV 25, 86, 116
swimming pool 47, 79, 87, 90, 118
Sydney 2, 94–5, 96–101, 213

taboos 42
tajine 18–19
take away 101
tatami flooring 49
television (TV) 17, 19, 23–5, 29, 46, 49, 52, 54, 56–7, 67–8, 71, 79, 81, 84, 88–9, 96, 101–3, 106–8, 116, 119, 122, 202, 214, 221–3
temple 37, 59
Tonga 2, 93–5, 105–8, 135, 142–4, 146, 148–50, 157–8, 161–3, 182–3, 198–9, 201, 211, 213–14, 216, 223–4, 230
top-down 10, 142, 208, 224
Toronto 2, 77–81, 213
tortoise (Tu'I Malila) 95
tourism 8, 37–9, 54, 149, 231
tourists 17–19, 37, 58–9, 82–3, 149, 228, 231
trade 3–5, 9, 224; blockage 148
traffic jams 114
train 47, 66, 70, 73, 143, 155, 181–2, 184, 194
trolleybus 170

Tucumán 2, 112–18, 175, 209, 213, 215
tukul 26–7

Uganda 26
Ulaanbaatar 2, 38–9, 53–7, 202, 213
Ume 47
umu 107–8
United Kingdom 2, 6, 8, 63, 70–4, 111, 135, 139–40, 143–4, 148–50, 154, 157–8, 165, 190, 193, 196, 198, 201, 211, 213, 216, 221–4, 230
United States 2, 5, 76, 78, 84–91, 135, 142–5, 148–50, 157–9, 163, 165, 168, 171, 173–4, 182–3, 190–6, 198, 201–2, 209–10, 213–14, 216, 221–4, 230
urban area definition 1–2; consumers 5, 152; employment 2, 4; middle class 36
ute 102, 105

van 51–2, 55, 154–5, 170, 172–3, 181
Vatican City 63
vegan(ism) 65–6
vegetarian(ism) 45, 65–6, 70, 72, 100, 148

Wackernagel and Rees 5–7, 126
walwal 32
waste 3, 5–6, 9, 43, 48, 71–2, 133, 189, 194, 197, 220; disposal 164; water 54, 68
water footprint 164
water wheel 188
well 21, 40, 85, 87, 102, 119, 212
well-being 228–9
wheat 157–9
White Nile 16
wildlife 9–10
wind (energy) 67, 69, 116, 131, 133, 180, 189, 194, 196–202; mills 188
wood *see* firewood
wood burner 57, 70, 87, 102
worship 11, 22, 188

yurt *see ger*

Taylor & Francis eBooks

Helping you to choose the right eBooks for your Library

Add Routledge titles to your library's digital collection today. Taylor and Francis ebooks contains over 50,000 titles in the Humanities, Social Sciences, Behavioural Sciences, Built Environment and Law.

Choose from a range of subject packages or create your own!

Benefits for you
- Free MARC records
- COUNTER-compliant usage statistics
- Flexible purchase and pricing options
- All titles DRM-free.

Benefits for your user
- Off-site, anytime access via Athens or referring URL
- Print or copy pages or chapters
- Full content search
- Bookmark, highlight and annotate text
- Access to thousands of pages of quality research at the click of a button.

REQUEST YOUR FREE INSTITUTIONAL TRIAL TODAY

Free Trials Available
We offer free trials to qualifying academic, corporate and government customers.

eCollections – Choose from over 30 subject eCollections, including:

Archaeology	Language Learning
Architecture	Law
Asian Studies	Literature
Business & Management	Media & Communication
Classical Studies	Middle East Studies
Construction	Music
Creative & Media Arts	Philosophy
Criminology & Criminal Justice	Planning
Economics	Politics
Education	Psychology & Mental Health
Energy	Religion
Engineering	Security
English Language & Linguistics	Social Work
Environment & Sustainability	Sociology
Geography	Sport
Health Studies	Theatre & Performance
History	Tourism, Hospitality & Events

For more information, pricing enquiries or to order a free trial, please contact your local sales team:
www.tandfebooks.com/page/sales

 Routledge — Taylor & Francis Group | The home of Routledge books

www.tandfebooks.com